Natural history of infectious disease

D0218696

Natural history of infectious disease

by SIR MACFARLANE BURNET

O.M., F.R.S., Nobel Laureate

and DAVID O. WHITE

Professor of Microbiology,
University of Melbourne

FOURTH EDITION

CAMBRIDGE UNIVERSITY PRESS

CAMBRIDGE

LONDON · NEW YORK · MELBOURNE

Published by the Syndics of the Cambridge University Press
The Pitt Building, Trumpington Street, Cambridge CB2 1RP
Bentley House, 200 Euston Road, London NW1 2DB
32 East 57th Street, New York, NY 10022, USA
296 Beaconsfield Parade, Middle Park, Melbourne 3206, Australia

Library of Congress Catalogue Card Number: 74–174264

ISBN: 0 521 08389 3 hard covers
ISBN: 0 521 09688 X paperback

First published 1940
Second edition 1953
Reprinted 1959
Third edition 1962
Fourth edition 1972
Reprinted 1974 1975 1978

First published under the title:
Biological Aspects of Infectious Disease

First printed in Great Britain
Reprinted in the United States of America

Contents

Preface   *page ix*

1 THE ECOLOGICAL POINT OF VIEW   1
   Historical approaches to infectious disease – general
   ecological concepts – host–parasite relationships – human
   ecology in relation to infectious disease

2 EVOLUTION OF INFECTION AND DEFENCE   22
   Origin of life and early evolution – self and not-self –
   evolution of infectious disease

3 BACTERIA   32
   Biochemical aspects – ecology and genetics – bacteria as
   parasites and pathogens – mycoplasmas, rickettsiae and
   chlamydiae

4 PROTOZOA   44
   Amoebic dysentery – trypanosomiasis – malaria

5 VIRUSES   52
   Laboratory manipulation – the ubiquity of viruses –
   their evolution – viruses pathogenic for man

6 INFECTION AND IMMUNITY   70
   Antibody and immunoglobulin: the T and B systems –
   recovery from infection – subclinical infection and immunity – prophylactic immunization – diagnostic applications

7 SUSCEPTIBILITY AND RESISTANCE   88
   Genetic differences – the influence of age – route and
   mode of infection

Contents

8   HOW INFECTIONS SPREAD   105
      Intestinal infections – the respiratory route – venereal disease – arthropod-borne infections – vertical transmission

9   EPIDEMICS AND PREVALENCES   118
      The determining factors – subclinical infection – shedding of organisms into the environment – survival outside the body – statistics of incidence – island epidemiology – the English Sweats

10   EVOLUTION AND SURVIVAL OF HOST AND PARASITE   137
      Impact of a 'new' infection – intrusion into an alien ecosystem: animal reservoirs – latent infections and vertical spread

11   CONTROL OF INFECTIOUS DISEASE   155
      Principles of control and prospects for eradication – quarantine – blocking the chain of transmission – protecting the individual – immunization

12   ANTIBIOTICS   173
      Sources and mode of action – resistance to antibiotics – allergy and toxicity – controlled therapeutic trials

13   HOSPITAL INFECTIONS AND IATROGENIC DISEASE   186
      'Hospital spread' of bacteria – infection by transfer in blood – immunization catastrophes – vulnerability of special groups of patients

14   DIPHTHERIA   193

15   INFLUENZA   202

16   TUBERCULOSIS   213

17   PLAGUE   225

18   MALARIA   232

19   YELLOW FEVER   242

Contents

20 HEPATITIS, KURU AND SLOW VIRUSES 250

21 PERILS AND POSSIBILITIES: AN EPILOGUE 262
Changes since 1940 – expectations for the future –
effects of major war – microbiological warfare – population control

Index 273

Preface

This book is a new edition, under joint authorship, of what first appeared in 1940 as *Biological Aspects of Infectious Disease* by F. M. Burnet, and has been revised at approximately ten-year intervals.

Substantial changes have been made in an attempt to bring the book up to date in a field that has moved at an astonishing rate over the last thirty years and even over the last ten. About half the book has been completely rewritten, and two new chapters have been added, on hospital infections and iatrogenic disease, and on hepatitis, kuru and slow viruses.

We have probably changed the character of the book significantly by assuming that the reader nowadays would have a rather wider general knowledge of biological matters than in earlier years and many of the more elementary dissertations of previous editions have been eliminated.

Nevertheless every effort has been made to retain the basic flavour and format of the original and it is still in no sense a text book. We have kept very much in mind the original objectives in writing it, without forgetting that the world has changed greatly in the interim. Those aims were, and are, first to stimulate scientific interest and provide an introduction of facts and ideas for young people contemplating a career in medicine or related spheres. The second objective is to present a consistently biological account that, hopefully, could be helpful to anyone, physician, scientist or layman, who has a real but peripheral interest in human infectious disease.

Much of the book has been modified only insofar as was necessary to update the factual content. Where a story first told in 1940 still illustrates a general principle just as satisfactorily as a more topical example would, it has been retained. Indeed, the senior author confesses to a measure of satisfaction in noting the surprising degree to which most of the basic principles of ecology he tried to enunciate over thirty years ago have lasted to this day. There is a certain permanency about the fundamental truths of a subject as close to nature as ecology. The rules of the game may not have changed for a billion years.

Ecology has suddenly become a fashionable word. Most thinking people everywhere have at last begun to recognize the importance of man's relationship to his global ecosystem – a realization that was long overdue. The senior author would like to feel that the first edition of the present book when published just over thirty years ago may have helped to define some of the concepts upon which the 'new' ecology is based.

Melbourne   F.M.B.
May 1971   D.O.W.

1.   The ecological point of view

In the final third of the twentieth century we of the affluent West are confronted with no lack of environmental, social, and political problems, but one of the immemorial hazards of human existence has gone. Young people today have had almost no experience of serious infectious disease. The classical pestilences, smallpox, plague, typhus and cholera have been banished effectively for a hundred years or more and in the last half century the standard childhood infections have progressively lost their power to kill. Scarlet fever, diphtheria and poliomyelitis have virtually disappeared and there is a clear promise in the air that measles will soon follow them into oblivion. A child in Australia, North America or Europe can still expect to experience a number of relatively trivial respiratory and febrile infections but the chance of his dying from infection before maturity is very small. For the first time in history deaths in infancy and childhood are not predominantly from infection. The killers of children nowadays are accidental mishap, congenital anomalies, or cancer.

From the beginnings of agriculture and urbanization till well into the present century infectious disease was the major overall cause of human mortality and the most important stabilizer of population levels. Now the whole pattern of human ecology has, temporarily at least, been changed. Of course the change is far from complete. Not one of the major plagues has yet been eradicated on a global scale, and if civilization collapsed and the machinery of public health could not be maintained they could swiftly return and play havoc in our crowded world. Infectious disease may be almost invisible but it is still potentially as important as ever it was.

In revising and rewriting this book we are conscious both of the virtual absence of serious infectious disease today and of the factors that made that consummation possible. We are equally aware of the changes this has wrought in human life and of the obscurity of a future which may bring new dimensions to infectious disease. More than most subjects the natural history of infectious disease must be set against a historical background and discussed in terms of continuing change.

1

Natural history of infectious disease

Over most of the historical period, the human attitude to epidemics and the other aspects of infectious disease was a curious mixture of erroneous theory with a good deal of useful common sense. On the practical side, we must remember that the contagiousness of some of the more easily recognized diseases like plague and leprosy was well understood in the Middle Ages, and logically planned measures to prevent or minimize epidemics were attempted. The institution of quarantine for travellers from infected regions by the Venetians in 1403 may be instanced, while a general recognition of the association of disease with civic filth and personal uncleanliness goes back to classical times.

Theoretical ideas took some fantastic forms, but after the time of Hippocrates there was nearly always evident a desire to replace the too simple explanation of disease as divine vengeance for wrongdoing by some more immediate cause. Infection seemed obviously to pass from person to person in the air, and it was a natural association of ideas to think of the spread of infection as analogous to the diffusion into the air of unpleasant smells from septic wounds or dead bodies. From the earliest times, putrefaction, especially of unburied human corpses, was regarded as likely to breed disease. In the early nineteenth century, 'bad drains', that is, smelly drains, were universally blamed for typhoid fever and diphtheria. Even as late as 1890, well-informed epidemiologists considered that epidemics of plague, influenza or yellow fever might be initiated by the diffusion of gaseous material from poisons in the soil, usually of putrefactive origin. They knew that a further generation of infective material in the bodies of those initially infected must occur and be responsible for spread of the epidemic, but they looked for the initiating cause of the epidemic in peculiarities of soil and weather.

With the coming of Pasteur and Koch, the mystery of infectious disease seemed to be swept away for ever. 'Fluxes, agues, botches and boils', not to mention the major pestilences, were all due to the attack of bacteria or similar microorganisms on the body. The new science of bacteriology provided a great impetus to the process of cleaning up the conditions of human existence which had begun, in England at least, a few decades earlier. A rational basis was now available for the demands for pure water supply, sewerage schemes, pure food and milk regulations, prevention of overcrowding, and so forth. Medical and particularly surgical techniques were revolutionized by antiseptic and aseptic methods. Typhoid fever disappeared, and infantile mortality rapidly declined, surgery free from the fear of sepsis flourished, and a few economically vital areas of the tropics were cleared of yellow fever and malaria. All these triumphs, and they were triumphs, received the

2

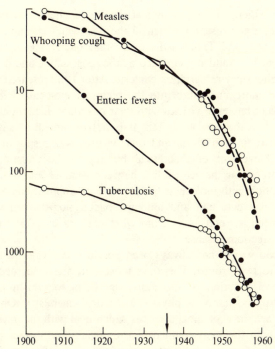

Fig. 1. To show the trend of mortality from some infectious diseases in England and Wales during the twentieth century. Relative death rates are shown for measles, whooping cough, enteric fevers and tuberculosis. The absolute scales differ but all are shown logarithmically, each marked interval indicating a tenfold difference. Between 1936, when the first edition of this book was being written, and 1960, there was almost one hundredfold diminution in the deaths from whooping cough and measles.

publicity they deserved. Medical research, and especially bacteriological research, became an eminently worthy occupation.

Along with this development of practical measures of preventing and curing infectious disease, there has gone a steadily increasing interest in the intrinsic characteristics of the microorganisms responsible for disease. In the early stages of laboratory research on infectious disease, most bacteriologists had been trained as medical men and their main interests were on the isolation of bacteria as causes of disease, the development of diagnostic tests and curative sera, and the experimental production of disease in the animals of the laboratory. Then as biochemistry developed, men with chemical training began to realize how convenient bacteria were for the study of many of the fundamental chemical problems of living substance. From the 1930s onward bacterial

3

physiology which, in fact, is almost wholly biochemistry, has been a favourite topic for research. Then in the post-war years and especially after Lederberg and Tatum's discovery that bacteria could hybridize, bacterial genetics suddenly emerged as the most active and illuminating aspect of laboratory research on bacteria. After 1960 research on viruses also moved more and more into the realm of molecular biology; the interactions of bacterial viruses with their hosts provided much that was fundamental to the new approach. It is not too much to say that the interlinked study of bacteria and their viruses was responsible for the greatest achievement of theoretical biology – the understanding of protein synthesis as the central link between genetics and biochemistry. The fascination with molecular biology and its implications at all levels of biology has persisted and amongst bacteriologists and virologists there has been a trend away from what may be called the ecological aspects of infectious disease.

In a broad way it has always been possible to divide the biological sciences into two groups. There are those that are concerned with the structure and function of the individual living organism and of its component parts, anatomy, physiology, and biochemistry. On the other hand there are the ecological sciences which deal with the interaction of organisms with their environment and especially with other organisms whether of their own or different species in that environment. Ecology is a broad term covering such an approach to the problems of animals, plants and microorganisms, and in the human sphere we can include the social sciences under the same heading. It is an achievement of the modern period to realize the existence of a third region, basic to both universes of study, within which belong genetics, evolutionary theory and the various aspects of embryology. It is in this zone that we find the growing edge of biology in the 1970s. In many ways the modern objective of both the classical biological sciences and the ecological sciences is to relate their findings to the concepts that are developing in this third bridging zone of biology.

Ecology however is what we shall be mainly concerned with in this book, and in its bearing on the natural history of infectious disease in man it becomes epidemiology. Everyone today uses the word ecology and has at least some grasp of the things with which it is concerned. But it is not something that can be given a simple definition or elucidated in a few sentences. A feeling for the general quality of the ecological approach is needed and this is what we attempt to provide in the rest of this chapter. We have deliberately retained the examples of ecological principles in action that were used in earlier editions. Those principles are timeless and it seemed wise to show that they were as relevant forty years ago as the world knows them to be today.

*

Since the eighteenth century, at least, there have always been educated men of some leisure with a natural interest in the activities of animals and plants. Many of these amateur naturalists, from Izaak Walton and Gilbert White onward, have written about the way animals make a living. The habits of birds in feeding, courting and nesting have attracted the interests of many. Others have spent years unravelling the life history of insects, especially the miniature civilizations of the ants, bees and termites. In more recent years, this essentially amateur type of observer has been supplemented by the professional biologist, whose more systematic investigations in the field once known as nature study have raised its dignity to the science of ecology. As the form of the word suggests, ecology is the study of the economics of living organisms. Animal ecology deals with the activity of animals as individuals and as species, their mode of feeding and of reproduction, the environmental conditions necessary for their well-being, and the enemies with which they have to contend. The combined action of such factors determines how numerous a species will be at any particular time and place. Sometimes inconspicuous changes in the environment of a species may produce extreme changes in its numbers, and in practice most ecological investigations are undertaken to elucidate the excessive abundance or undue scarcity of some economically important animal. Like most, perhaps all sciences, ecology has developed in response to practical human needs. On the one hand, there are countless pests which, when their numbers are excessive, cause great economic loss from damage to growing crops, domestic animals or stored food products. On the other hand, there are the economically important wild animals, the fur-bearing rodents and carnivores of subarctic regions, the edible fish of the North Sea, the whales of the Antarctic, and so on, where the problem is to avoid undue scarcity or extinction of the species. If the pests are to be controlled or the valuable species saved from extermination, every detail of their life histories, their physical environments, and of the numbers and habits of their enemies may be necessary. It is the task of the trained ecologist to provide this knowledge and to show how it can be applied to the desired end.

Every animal species reproduces its kind at a far greater rate than would be necessary to maintain its numbers if death occurred only as a result of old age or accident. In nature, of course, limiting factors are always present, and the maintenance of the numbers of a species can be regarded as the result of conflict between two opposing forces, the 'population pressure' of the species, the constant production of more young than can hope to survive, and the equally constant destruction of individuals by physical calamities, frosts, floods and the like, predatory and parasitic animal enemies and infectious disease. Except for some large, slowly breeding animals, the reproductive potentialities of most species are so great that it needs only a surprisingly short time for a vast increase in numbers to occur following some change in the environment unusually favourable to the species concerned. In most practical problems, this capacity for rapid multiplication can be taken for granted, and attention need be focused only on the two principal factors which diminish or limit the numbers of any species – the available food supply and the activity of enemies. The animal enemies of a species include those which capture and devour their prey (predators) and those smaller forms which live in or on the tissues of their host (parasites); and not infrequently there are enemies whose harmful activities are intermediate between those of predator and parasite – the blood-sucking insects and vampire bats for example. All these activities of enemies, however, have the same object, to obtain food for their own requirements from the tissues of their victims. Put concisely, the two essentials for survival are that an animal should find enough to eat and avoid being eaten itself. Food thus becomes the central problem of ecology.

Every animal needs for its nutrition some form of protein which can be used to build up its tissues, and carbohydrate or fat to provide energy for its activities. In the body the chemical form of these substances can be modified, but the essential 'building-stones', amino acids and simple carbohydrates, must be obtained from without. Only the green plant can synthesize these substances from inorganic materials, and all animals are primarily dependent on plants for their nutrition. Probably more than half the species of animals and the vast majority of individuals feed directly on plants or their products. There is an infinite variety in the methods of such feeding. The single-celled protozoa engulf minute green algae, fish and molluscs browse on seaweed, insects and humming birds seek honey in flowers; there are eaters of seeds and fruit, leaves, bark and roots; insect larvae and termites tunnel through wood, using it as food in the process, and there are multitudes of insects which exist by sucking the juices of plants. A few species of animal, protozoa and certain corals, have dispensed with the necessity of devouring plant material by incorporating small green algae in their tissues and utilizing the food material synthesized by these primitive plants.

All these plant-feeding animals by their digestive processes break down the plant substances to simpler molecules, and discarding waste materials build these up into their own proteins and fats. In doing so they provide for other creatures a more concentrated and more readily utilized store of food. Nearly every animal is liable to become the prey of some predatory carnivore, and with the exception of the largest carnivores, this holds irrespective of whether the animal itself feeds on plants or on other animals. For every animal species one can trace back its food supply through one or more stages to the ultimate source in plant tissues. To take an unusually complicated example, a shark feeds on large fish which in most cases capture smaller fish. These probably find their main food supply in small crustaceans which feed on the protozoa of the surface waters. The protozoa live on microscopic green algae, the unicellular plant organisms which are the final source of food for most marine animals. Such a sequence of organisms, each feeding on the one beneath it, is referred to as a food-chain. Amongst land animals the food-chains are usually composed of fewer links than amongst the larger marine forms. A lion feeds on plant tissues at only one remove, since large herbivorous mammals are its usual prey. A longer chain leads from the small but very numerous insects, aphids, plant lice, scale insects and so on, which suck plant juices. These are preyed on by the larvae of ladybird beetles, which may provide food for birds either directly or after being eaten by spiders. Owls and hawks, by feeding on the smaller birds, complete the chain.

In all the series we have mentioned so far, the animal which eats is larger than that eaten. At the end of each food-chain we find the larger carnivores, eagle and owl, wolf and lion, killer whale and shark, along with a few herbivores like the elephant, which, by their size or for some other reason, are immune from attack. Such animals have no visible enemies in nature, but they are just as exposed to the attacks of parasites as smaller types. The tiger may be the lord of the jungle, but its lungs may be riddled with parasitic worms.

The parasitic mode of life is essentially similar to that of the predatory carnivores. It is just another method of obtaining food from the tissues of living animals, and it is sometimes not easy to decide whether a given form is or is not a parasite. In general, a parasite may be defined as an organism smaller and less highly differentiated than its host, which lives on the skin or within the tissues or body cavities and gains its nourishment at the expense of the host's living substance. Although at present we are concerned with the means by which animals gain their food, we can class all the internal parasites together. The microorganisms of disease, just as much as parasitic worms, are using the tissues of their host simply as a source of the food they require for growth and multiplication.

As an example of the way in which this complex interaction of species feeding on and forming the prey of other organisms is reflected in the varying numbers of the species concerned, we may describe one of the first and most striking successes of applied ecology. Soon after orange growing had become an important industry in California, serious losses began to occur through the spread of a scale insect. The trees were covered with little white cushions built up by the insects as a protective covering beneath which they sucked the sap of the orange trees. It was soon found that the pest was not a native American insect, but an importation from Australia. There it normally lives on the native acacias, and when it spreads to orange trees does no particular damage. The difference between the behaviour of the insect in Australia and in California was not due to climate, but to the fact that in its native home the cushion scale insect is the chief food supply of a ladybird. The two insects automatically control each other's numbers. If the scale insect is particularly plentiful, the ladybird larvae find an abundant food supply and the beetles in their turn become more plentiful. An excessive number of ladybirds will so diminish the population of scale insects that there will be insufficient food for the next generation, and therefore fewer lady-birds. On the whole, a balance will be reached with such a population of each species that the destruction by enemies is approximately equal to the increase in numbers allowed by the available food supplies.

In California there were no carnivorous insects adapted like the Australian ladybird to feed on and reduce the numbers of scale insects to a normal level. The logical way to remedy the excessive multiplication of the pest was to introduce its most important natural enemy into the orange groves. This was done in 1889, and once the difficulties of breeding a sufficient supply of the little beetles had been overcome, the experiment was a triumphant success. The cushion scale was reduced in importance to a relatively trivial pest.

In this example we have a contrast between the behaviour of an organism, the scale insect, in its natural environment and in a new environment, which in essentials may be taken as the prototype of a large number of practical ecological problems. In Australia, the scale insect, like the other indigenous plants and animals, had been free to evolve and adapt itself to a relatively constant environment for millions of years. In the long period before extensive European settlement, the mutual action of all the species, animal and plant, had determined what numbers of individuals of each species would allow an approximately balanced and stable condition to be maintained. The interaction between scale insect and ladybird is only one of countless similar adjustments between all groups of species whose food-seeking activities influenced one another's numbers.

The process by which an approximately constant population of living organisms is developed has been studied in more detail by botanists than by animal ecologists, but the broad principles are the same for both great classes of living organisms. The final constant condition of the vegetation in any particular region is called by plant ecologists the 'climax' state. It is not infrequent for the normal vegetation of an area to be completely destroyed by human activity or as a result of some natural catastrophe, fire, flood or volcanic eruption. If after the destruction the land is left to natural processes, a regular series of changes can be recognized in the return of plant life to the area. The island of Krakatoa in Indonesia provides the most striking example. In 1883 it was apparently completely sterilized of every living organism by intense volcanic activity. Within sixty years it had again become covered with dense tropical forest. First of all the lifeless volcanic ash was colonized by air-borne spores of bacteria, primitive plants, mosses and ferns, then in succession appeared grasses and shrubs from sea-borne seeds. Twenty-three years after the eruption, a forest of coconut palms, fig trees and others had developed along the shore and was gradually spreading inland. Today the tropical rain forest has taken over the areas not involved in the subsequent minor eruptions of 1927 and 1953. This succession of different types of vegetation is seen in all such denuded areas left to uncontrolled natural processes. In every case the succession goes on to a final state characteristic of the region, and determined mainly by the climate and nature of the soil. This final 'climax' may be tropical rain-forest, grass-covered prairie, pine forest or some other type of plant community, but once reached, it remains essentially constant, and is likely to change only as a result of long-period climatic changes or human activity.

Along with such changes in the plant life, a related series of changes in the animal population will occur, and with the establishment of the climax state of vegetation animal species will adjust themselves to an uneasy equilibrium. This mutual adjustment is an immensely complicated process, for all the food-chains concerned are naturally interwoven, and for every species there will be fluctuations in numbers from time to time, but on the whole, in a constant environment a reasonable

9

approach to a stable balance will be maintained. As long as the number of individuals does not fall below a certain minimum, the lower the population density of a species, the greater the opportunity it has to multiply. There is a greater food supply available for each individual; general predatory enemies may consume much the same proportions, but there is less opportunity for enemies such as parasites and carnivores of restricted prey to thrive at their expense. Conversely, when the numbers increase, food supplies are limited, and there is increased opportunity for the activity of all those specialized parasites and predators that are more or less strictly confined to the species in question. So, from each side, there is an automatic tendency to the development of a balanced condition of affairs for every species. It is an uneasy balance, more often swinging widely above and below the equilibrium point than remaining nearly stable, and liable to be forced to a totally different equilibrium by the introduction of some new organism into the environment.

Australia provides some particularly striking instances of such changes, probably because the whole continent was practically isolated from the rest of the world for the long geological period from the Eocene until the European settlement. Plants which are relatively unimportant pests in their native habitats have only too often spread excessively after their introduction into Australia and become dominant forms in certain areas. The prickly pear cactus is the best known, but in the south the blackberry will become the dominant vegetation of fertile valleys unless kept continuously in check by human action, and thistles have spread intensely over the plains country. Amongst animals, the rabbit, hare and fox have almost completely replaced the native mammalian fauna in the south-eastern region. Only the arboreal opossums have managed to maintain their numbers.

Australia in its turn has provided pests for other countries, mostly insects. We have mentioned the cottony cushion scale of acacias and orange trees already. A leaf-hopper from Queensland at one period almost destroyed the sugar industry in Hawaii until suitable parasitic wasps from Australia were introduced to control its numbers. There are dozens of similar examples to be found in the literature on the spread and control of economically important pests, and there must be an infinite number of similar adjustments going on amongst organisms not important enough to man to call for detailed research.

This development of an approximately balanced condition between contending species is as characteristic of the relation between host and parasites as of any other such interaction, but there are certain special characteristics of the parasite–host relation which need to be considered. Most parasites are restricted to one host species, or at most to a small number of related species, and the main problem that a parasitic species has to solve if it is to survive is to manage the transfer of its offspring from one individual host to another. All sorts of methods have been developed: the production of enormous numbers of offspring is usual, and many larval forms find an intermediate host in some other animal whose movements or activities will help the transfer to fresh, final hosts. It will be clear, however, that no matter by what method a parasite passes from host to host, an increased density of the susceptible population will facilitate its spread from infected to uninfected individuals. Suppose we have ten fowls in a large enclosure, one of them infected with coccidia, a protozoal parasite of the bowel which is spread by contamination of food with infected droppings. It will probably be a long time before another bird becomes infected, and when it does, only a small number of parasites will be taken in and the fowl will probably escape with a mild infection. But if 500 fowls are placed in the same enclosure, again with one infected individual amongst them, there will be a fifty times greater chance that another bird will be infected within a short period. As soon as a few become infected and start excreting the parasite, the process will spread with accelerating speed until all are infected. Further, they will be constantly subject to reinfection, and liable to more severe symptoms and a higher mortality. This is precisely what is found to happen in poultry-farm experience if precautions against coccidial infection are not taken.

As a more natural example of the mutual influence of population density and parasitic activity, we may say something about the plagues of small rodents which occur periodically. The lemming swarms of Scandinavia are the classical examples, but in Australia we are more familiar with mouse plagues. In certain years the introduced domestic mouse multiplies enormously. The mice swarm in crops and haystacks, and literal bucketfuls can be caught in a single night. Hawks, owls and cats flourish at their expense, and birds like ibises change their normal diet to one of mice, but all these enemies have little effect in reducing the numbers. As a rule the plague ends rather suddenly. A few dead mice are found on the ground and the numbers dwindle rapidly to, or below, normal. Infectious disease has been at work, and when the population reaches a certain density, rapid and concentrated transfer of parasites, bacterial and otherwise, takes place. Not many opportunities have arisen for bacteriologists to determine what are the actual microorganisms responsible for these epidemics which bring mouse plagues in Australia to an end. We can be sure that the type of parasite will vary from one occasion to another.

In 1940 for instance the wheat-growing areas of Northern Victoria had been suffering from a typical mouse plague for some months when mice were sent down to Melbourne for bacteriological examination. All the mice were diseased with a fungal infection of the skin of the face, and large abscess-like swellings on the limbs due to an unusual type of bacterium (*Streptobacillus moniliformis*). On another occasion a mouse plague in Queensland was associated with widespread human cases of a form of typhus fever, and it is reasonable to think that it was an infection of this sort amongst the mice which terminated the period of over-population.

Such a reduction in numbers by the activity of parasites (including the microorganisms of disease) when animal populations become abnormally large is only another manifestation of the principle that for every organism there is a certain density of population which makes best for survival of the species.

*

How far is this ecological viewpoint applicable to man? For centuries he has been securely at the right end of a whole multitude of food-chains. But, like other mammals, he has persistently multiplied beyond the limits of subsistence, and in the absence of man-eating carnivores, other checks on population growth have developed. Simple starvation may sometimes have been important, but more often shortage of food supply has had indirect effects in forced migrations, internal and external war, diminished fertility and excessive infantile mortality. All of these have helped to effect the needed reduction in numbers. Even more important has been the influence of famine in predisposing to increased prevalence and increased intensity of infectious disease. War and famine have probably always taken their toll of human life more through the intermediary of the microbe than by starvation and the sword. And even in those rare phases in history when peace accompanied prosperity, infectious disease has played its part. Right up to the nineteenth century infectious disease was the most important agent in preventing human overpopulation, and in a few parts of the world it still remains so. With its progressive elimination from most centres of human population, infantile and childhood deaths are falling steeply and the greatest political problem of the second half of the twentieth century is now taking shape.

It needs only the most limited knowledge of human history to see that the whole process of civilization has been largely one of aggregation of greater and greater numbers of human beings into limited areas. Cities are essential to civilization, and until well into the nineteenth century, all cities were the spawning grounds of infectious disease. How could it be otherwise? For centuries all the precautions we now know to be necessary to prevent the spread of bacteria and animal parasites were unthought of. City streets were littered with human and animal filth, water came from contaminated wells, rats, fleas and lice were universal.

Crowded together in such filthy environments, every city dweller was inevitably exposed to infection every day of his life. It is no wonder that the population of cities through all history has had to be recruited periodically from the country. Few cities were ever able to maintain their population by their reproduction, but the attraction of gregarious life has always been sufficient to bring a constant stream of the ambitious from the healthier, because less densely populated, countryside. And on a larger scale we see almost the whole of ancient and medieval history dominated by the periodic outpouring of nomadic tribes from the steppes and mountains of eastern Europe and central Asia down to the areas of civilization and cities in China, India, Mesopotamia, the Mediterranean basin and western Europe. The nomadic life is the healthy life, and the children of nomads survived until their numbers were too great for their steppes to support them. After each outpouring, the conquerors adopted the city life of the people they dispossessed, and disease saw to it that no longer did their children survive in the numbers that had been theirs in the nomad existence.

It will be our task in subsequent chapters to discuss the reasons why *all* dwellers in the unsanitary cities did not die of infectious disease. Here we are looking only at the general ecological problem of human existence in crowded communities. We might reasonably expect that if a densely populated area remained isolated from all other populous regions, an equilibrium condition would eventually be developed, irrespective of what parasites were initially present. It is equally reasonable to expect that the introduction of some important new parasite into the area would seriously disturb this equilibrium. Human history gives many examples of such conditions. Prior to the European discoveries, many of the Pacific Islands were quite densely populated. There were a few endemic diseases, but, on the whole, the people were very healthy. When the trader, the missionary and the blackbirder had finished their work, epidemics of a dozen or more infectious diseases of European origin, plus general demoralization, had reduced the population of such groups as the New Hebrides to a fraction of the original number.

In the European cities, with which epidemiological history mostly deals, there was always some degree of contact with other cities. New parasites might creep like the Black Death from city to city with the rats and their fleas, enter them with conquering armies as syphilis did in its sweep through Europe at the end of the fifteenth century, or spread invisibly from carriers like the cholera 120 years ago. Sieges, floods or drought brought famines to make other violent disturbances of the balance between man and his parasites. There was never an opportunity to develop a *modus vivendi* with all the microbes and larger parasites which assailed him.

Suppose we leave the insanitary past and survey the conditions of present-day civilization from the same point of view. Several things have happened. First, we have found methods of preventing certain types of infection which were of the greatest importance in the old days, those spread by filth, or, to be more direct, by human faeces, and those transmitted by animal parasites or semi-parasites such as fleas, lice and mosquitoes. Efficient sewerage and water supply, plus ordinarily decent cleanliness, have thus rid us in temperate climates of typhoid, dysentery, cholera, plague, typhus and malaria. We have not been able, and perhaps never will be able, to block the spread of those diseases which are transmitted by what is technically called 'droplet infection'. Colds, sore throats, influenza, measles and the like are all passed from person to person via tiny drops of saliva which are sprayed into the air during coughing, shouting and so forth. As long as human beings in large communities have to go about their business, such dissemination of infection will continue. Indeed we could well forecast an actual increase in the incidence of droplet-borne infections as the cities fuse into 'megalopolises' where the numbers of people in potential contact, in the epidemiological sense, may reach 100 million or more. Control will probably only be attempted when some particular type of disease becomes an evident danger or some new technique of prevention is discovered. Immunization will probably remain the standard approach to control with chemoprophylaxis (protection by antibiotics or other drugs) held in reserve for emergencies.

It is an interesting reflection of the changes which have taken place in the last thirty years that in the earlier editions of this book it was forecast that natural forces would lead to a state of virtual equilibrium of the common childhood diseases with a small and probably diminishing toll of death and disability. Today we are not content to leave things to Nature. There are effective methods of immunization against diphtheria, polio, whooping cough and measles while recently developed vaccines will probably soon bring mumps and rubella under equally effective control. More and more the ecology of infectious disease is being dominated by deliberate human action designed to provide at least short-term protection for the individual. It is a policy that may have some unhappy long-term repercussions but it sets the stage for the future pattern of infectious disease.

Ecology has become one of the vogue words of the 1970s and we are all talking of the disastrous disturbances of natural ecosystems that the apparently irresistible pressure of technological advance has produced. It is almost an axiom that action for short-term human benefit will sooner or later bring long-term ecological or social problems which demand unacceptable effort and expense for their solution. Nature has always seemed to be working for a climax state, a provisionally stable ecosystem, reached by natural forces, and when we attempt to remould any such ecosystem we must remember that Nature is working against us. Perhaps this may not hold to the same extent for our activities in preventive medicine but minor and major difficulties and disappointments have been common in the past and will no doubt arise in the future. It will always be desirable to understand the natural equilibrium, the more or less stable ecosystem, that determines the form of any infectious disease of man in the absence of deliberate human interference. Only with such a background is it possible to judge the practicability of proposed measures of prevention.

Let us take measles as an example of a typical childhood disease in which a single attack is followed by lifelong immunity. Of all infections its ecology is the simplest to understand. Its incidence is characteristically in epidemics separated by periods when measles is absent or at a very low level. We can picture the sequence by starting with the situation in a city where the disease has been absent for a year. A child becomes infected, perhaps from someone from another city. From this child a number of five-, six-, and seven-year-olds are infected, perhaps at school, and a fairly rapid spread occurs to all exposed children not yet insusceptible by virtue of a previous attack. This will mean, as a rule, children who have just commenced school and younger pre-school children in the same homes. Almost every susceptible child exposed to contact with another child in the early stages of infection will contract the disease. There may be some children who come into contact with infection but receive so small a dose of the virus that no real infection follows. Those children who are infected and have a typical attack of measles will practically all recover completely if they were well nourished and healthy at the time of onset. Amongst them however there may be occasional deaths – from septic complications or from encephalitis and some will suffer sequelae from non-fatal encephalitis. The epidemic will spread progressively through a city or a country, twelve or fourteen days between successive crops of infections, and under ordinary city conditions will not diminish greatly until a high proportion of susceptible children is infected. As each passes through his infection, he becomes immune and the virus finds progressively greater difficulty in spreading. Eventually measles disappears from the community, but in the meantime the virus is spreading similarly through other communities. In about two years' time another crop of susceptible children has appeared, and sooner or later the measles virus reenters the community and the cycle is repeated.

In a community where nearly all infants are immunized with a satisfactory measles vaccine the new crop of susceptible children is no longer available. Already there is good evidence that this great reduction in the number of susceptibles not only means that the immunized children do not get measles but also protects the small proportion who were not immunized. For an epidemic to develop there must be adequate opportunities of contact between infected and susceptible children. When the proportion of susceptibles is below a certain level measles must vanish.

The exact opposite of this situation was seen when measles came into the Pacific in the nineteenth century. The first epidemic in Fiji is probably the most interesting. In 1875 a son of one of the native rulers had been on a visit to Sydney and returned just as he began to show symptoms of measles. Epidemiologically it was a most unfortunate home-coming. It had been timed to allow the prince to take part in a major meeting of chiefs and dignitaries from all parts of the islands to mark the British Government's acceptance of Fiji as a Crown Colony. The festivities lasted long enough for measles to take a firm hold and to be transported immediately to every centre of population in the new colony. According to a modern demographer the mortality was probably about 20 per cent of the population. Part of this excessively high death-rate was probably due to the fact that the population had never previously experienced the disease but other factors may have been equally important. No one was immune and in many villages such a high proportion were ill simultaneously that social life broke down completely, there was intense fear and depression, inadequate food and no one to care for the sick. Under the circumstances one can also be certain that secondary bacterial infections played a part in producing the high mortality. After the first onslaught measles epidemics in the more populous islands were not greatly different from what was seen in European populations.

The lesson to be learnt from measles then is that with a virus of high capacity to spread and immunizing solidly against subsequent infection, the characteristic manifestation will be regular epidemics involving children at the earliest age at which they come together in groups. The time between epidemics will depend mainly on the size of the community and the ease of movement to and from other communities. Measles cannot persist indefinitely in small isolated communities. That statement carries the very important implication that in principle vaccination against measles could allow the eradication of measles from the globe.

Our second example to illustrate ecological aspects of the interaction of host and parasite takes a completely different form. Psittacosis or parrot fever is a severe disease with a high mortality in elderly people, known since 1890 to be contracted from parrots. In 1929–30 psittacosis became a matter of international importance owing to the large number of human infections which were derived both in Europe and North America from infected parrots imported from South America. Like many other infectious diseases, psittacosis was first recognized as a serious epidemic disease of human beings, but as its nature became gradually understood, it grew clear that the epidemic phase was only an accidental and relatively unusual happening. It is worth while going into the story of the elucidation of psittacosis, for it is an excellent example of the way interest in infections tended to shift from the medical to the biological aspects. The scientific study of psittacosis began early in 1930, when research workers in England, Germany and America almost simultane-ously announced that the disease was due to a small microorganism which we now classify as a *Chlamydia*. At once there became available ways of defining the disease and of isolating the causative organism from a patient or from a parrot. Only with facts so obtained was it possible to seek an understanding of the natural history of psittacosis and three years later the first important steps were made in the budgerigar aviaries of California.

In the United States, after the epidemic of 1929–30, there was a rigid prohibition on the importation of South American parrots, and it was hoped that this would be sufficient to get rid of the disease. But now that the disease had become familiar to doctors, cases kept on being recognized, and most of these had no connexion whatever with imported parrots. The sources of their infections were nearly all American-bred budgerigars. There was one relatively large outbreak traced to a consignment of these birds from a Californian aviary, and a large-scale investigation of the conditions in the budgerigar breeding establishments of the States was set going by the Californian authorities.

Most people know the budgerigar, also called the shell parakeet or lovebird. It is a native of Australia, very common in the sparsely timbered grasslands of the interior. The wild bird is predominantly green in colour, but under domestication various colour varieties have appeared and been developed by selective breeding. Now there are varieties of every colour, from nearly white to deep blue, with dozens of blends of yellow, green and grey. In most parts of the world it comes a close second to the canary·as the most popular cage bird, and there is quite a large minor industry devoted to their breeding and sale. In America, buderigar breeding on a substantial scale was almost confined to California, from which large numbers of birds were exported to the other States.

In the Californian aviaries there was ample opportunity to observe the behaviour of psittacosis. Over half of the 104 aviaries examined were infected. In many of these it was known that a considerable proportion of young birds became sick, and that some died, but others, also infected, had no undue mortality amongst their birds. Meyer, who was in charge of the investigation, came to the conclusion that in the infected aviaries most of the fledglings contracted the disease from droppings of older birds. The infection might or might not result in visible symptoms, but in any case most birds recovered. By the time they were eight months old, the budgerigars were free from all signs of the infection. If it was present at all, the organism now remained only in very small amounts in the cells of the spleen or kidney. Enough of these carriers persisted, however, to infect the young at the next breeding season. On the basis of these results, measures were set going to eliminate the disease, and a very considerable success was achieved.

In the course of this work, Meyer imported 200 wild budgerigars from Australia, assuming that these would be sure to be free from infection. A month or two after arrival in America, a few of them were noticed to be sick, and it was found that they were infected with psittacosis. They had had no known opportunity to become infected in California, and it seemed that they must have brought the organism from Australia. It obviously became necessary that Australian microbiologists should look into the conditions in their own land, which had heretofore been assumed free from the disease. In Melbourne we undertook to make the necessary investigations. As a preliminary, a batch of young parrots was purchased from the nearest bird shop. About a third of them were infected with a mild, but perfectly definite psittacosis. Subsequent investigations showed that most of the common Australian parrots and cockatoos were liable to be infected in the wild state. Sometimes almost every bird in the batches received direct from the catchers was infected. Other lots showed only a few with signs of past infection, and none actually carrying the organism. Each of the three main families-cockatoos, parakeets, and lories-provided its quota of infected birds.

At the time these investigations were made we were inclined to look on the 'virus' of psittacosis as something that had evolved with the parrots in Australia and had spread with the budgerigar to the rest of the world. Subsequent years, however, saw a steady widening of the range of birds known to be infected with agents of the same general character as psittacosis. In the Faroes, in the North Atlantic, cases of severe pneumonia began to occur around 1933, almost always in women and at the time of the year when large numbers of young fulmar petrels were being killed and prepared for food. A few years later it was established that the women were being infected by psittacosis, that the infection was widespread amongst the fulmars and that the young petrels on the bird cliffs of the Faroes were as commonly infected as the nestling budgerigars in the aviaries. In America, England, Australia and South Africa the semi-domesticated pigeons of the cities have been found carrying a mild form of the organism. Hens and ducks and several species of sea birds have also been added to the list, and it is hard to avoid the belief that we are dealing with a very ancient, almost universal infection of birds.

The conditions in Australian parrots may, however, still be taken as representative of the way these organisms survive in nature. It is not practicable to carry out the same detailed investigation of wild birds as is possible with aviary budgerigars, but our results were just what would have been expected if conditions in the wild were much the same as in the aviaries of California. The nestlings in all probability become infected from their parents, suffer a mild attack, and rapidly recover, but carry the organism in their bodies for at least a year. In this example, then, we have a well-balanced, mutually successful interaction between parasite and host. In the natural state it seems that very few parrots die or are even discommoded by the infection.

We may round off the story of psittacosis with an account of a small epidemic in Melbourne which we investigated over thirty years ago. It is an instructive little tale. In Australia, white cockatoos are popular pets; they are good talkers, and can at least appear to be both intelligent and affectionate. The demand is supplied by bird-catchers, who take the young from the nests in hollow trees just before they are able to fly. In the city, the birds are sold in animal shops and by casual dealers. The epidemic with which we are concerned arose from a batch of cockatoos kept for some weeks by a hawker in very confined and dirty quarters. Two serious human cases of psittacosis, one in the wife of the hawker, were diagnosed, and an investigation of the circumstances was made by the health authorities. It was found, first, that all the remaining cockatoos in the backyard shed were sick and heavily infected with psittacosis; and secondly, that altogether a total of fourteen people had probably been infected from this batch of cockatoos. Only five of the cases, however, were severe, and there were no deaths.

Here we return to the medical sphere: something had tipped the balance in favour of the parasite. It is reasonably certain that those cockatoos, left to a natural life in the wild, would never have shown any symptoms of their infection. In captivity, crowded, filthy and without exercise or sunlight, a flare-up of any latent infection was only to be expected. The organism multiplied and spread throughout the body. Large amounts of it passed out with the excreta, soiling the feathers. When the birds fluttered they were always liable to fluff up a cloud of contaminated dust and infect anyone in the neighbourhood.

Psittacosis is not intrinsically a very infectious disease. It is very rare for infection to pass from one human being to another. To produce dangerous illness the agent must be inhaled into the lungs, and ordinarily this is likely to occur only when infected parrots or cockatoos scatter infectious material into the air by their movements. In one or two instances, sick parrots brought to a laboratory have infected people in all parts of the building by the spread of such dust on air currents along the corridors.

Our Melbourne epidemic was only a small one, but it created a good deal of public interest, pointed a moral against the iniquities of bird-catching, and for the first time made it clear that the mild natural psittacosis of our wild cockatoos could light up to an intensity capable of infecting man.

Perhaps these examples are sufficient to indicate the general point of view which we must adopt in regard to infectious disease. It is a conflict between man and his parasites which, in a constant environment, would tend to result in a virtual equilibrium, a climax state, in which both species would survive indefinitely. Man, however, lives in an environment constantly being changed by his own activities, and few of his diseases have attained such an equilibrium. The practical problems of prevention and treatment demand principally an understanding of the results of new types of infection on the individual and the community, and of the stages by which a condition of tolerance is reached. When such knowledge has been gained it is a simple matter to apply it to the interpretation of those diseases which have reached their climax state.

Two interdependent sciences have arisen to deal with the problems of infection. Immunology is concerned with the response of the *individual* to invasion by microorganisms. What determines whether he lives or dies, what are the physiological processes by which the disease is overcome, and what is the basis of the immunity to further infection which so often follows the attack? Epidemiology deals with the large-scale phenomena of infectious disease, not only with epidemics but with the less dramatic, more or less constant, prevalence of disease in human communities. Its concern is with the *community* or the human race as a whole, not with the individual, but it is self-evident that its methods and generalizations must be based largely on what is known about the individual's reaction to infection. It is the aim of this book to present a simple account of the microorganisms that cause disease, of the processes within the body that are called into action against them, and of the way infection persists and spreads within the community. It is written not from the point of view of a physician or a pathologist, whose interests must necessarily centre on the human aspect, but with the outlook of a biologist, to whom both man and microorganism are objects of equal interest.

2. Evolution of infection and defence

In studying the nature of disease we cannot restrict ourselves to the confines of human medicine. The whole range of living beings comes into our province, for there is probably no species of organism which has not at some time been either host to a parasite or a parasite itself. Many have filled both roles. Infectious disease is universal, and any attempt to imagine how it arose in the course of evolution will inevitably take us back to the very earliest phases of life.

The astronomers and geologists have now given us a fairly detailed picture of the origin of the solar system and of the gradual development of the surface characteristics of the earth. Most authorities now believe the earth originated in the complex process by which a huge disc of what can be called cosmic dust aggregated to form the sun. A solid earth has existed for about $4\frac{1}{2}$ thousand million years. Its original atmosphere, mostly of hydrogen, was lost at an early stage and as the earth consolidated and heated up a secondary atmosphere was produced from the volatile materials 'squeezed' from its substance – water, ammonia, methane, hydrogen, carbon monoxide and nitrogen, but no free oxygen. As cooling progressed, the water vapour liquefied and formed the earliest oceans. Up to this stage there was no possibility of life. We can be quite certain that no living organism existed on the earth at the time of its formation, and almost as certain that by no conceivable process could life reach the earth preformed from any part of the universe. Living organisms must at some time have arisen from non-living material on the earth.

The orthodox conjecture is that the early oceans steadily accumulated a wide variety of organic compounds formed from gases under the influence of lightning and the ultraviolet radiation which penetrated the primitive atmosphere much more effectively than it does today. Laboratory work on similar gas mixtures exposed to repeated electric discharges has shown that amongst the simple organic compounds produced are most of the building blocks from which the giant molecules characteristic of living substance are constructed. Such building blocks dissolved in the water of primitive seas became the raw material from which time and circumstance allowed life to develop. Bernal has suggested that in the

mud banks intermittently exposed at the edge of shallow seas the conditions were appropriate for the appearance of an enormous variety of organic compounds. Here there would be opportunity for exposure to ultraviolet light and also for protection of unstable compounds from undue exposure, as well as for adsorption of substances to the reactive surface of clay particles. In some such situations the first precursors of living matter may have appeared.

Steadily increasing knowledge about the chemical structure of the smallest organisms has stimulated some detailed speculations about the nature of these intermediates between not-living and living matter. Space travel today makes it possible to study matter from planets other than our own and is bringing to fruition the biologist's dream of transporting the problem of the origins of life right into the laboratory. It now seems unlikely that there is life on Mars but if there is it will need only a handful of surface soil with its contained microorganisms to keep biochemical laboratories throughout the world busy for years sorting out the common features of the emergence of living matter on two separate planets. However, any discussion of the possibilities would be quite outside our present scope. All that need be said is that in one way or another giant molecules or the crude beginnings of organisms commenced the process of incorporating dissolved compounds of carbon and other elements into growing self-persisting units. The essence of life is this quality of incessant incorporation of chemically suitable material into living substance which, in its turn, continues the process.

We can be sure that at some point in evolution a certain standard chemical structure appeared which was so supremely successful that it alone survived. There is a basic uniformity of chemical structure to be found in the whole range of living organisms from virus or bacterium to man. If there were once organisms made to some other pattern, they were tested for survival and found wanting in competition with the standard pattern of at least the last 1,000 million years.

Today we find that all forms of life make use of essentially the same biochemical mechanisms for carrying out the fundamental life processes, of passing on inherited information to daughter cells and translating that information into the proteins that enable the cell to function. All living things utilize a single standard 'genetic code', whereby a given sequence of nucleotides in the nucleic acid that comprises a gene determines a correct sequence of amino acids in the corresponding protein molecule for which that gene codes. For a masterly description of this and all other aspects of the new discipline of molecular biology, representing the modern fusion of genetics and biochemistry, the reader is referred to

Watson's *Molecular Biology of the Gene*. Discussion of how these genetic and biochemical mechanisms evolved is a fascinating pastime for those with technical expertise and a taste for speculation but it is not really relevant to our theme of the evolution of infectious disease in animals and man.

Of all the possible voyages in an imaginary 'Time Machine', a biochemist would probably find a journey to the age when these first developments of living matter were occurring the most enthralling. One can imagine strange changes taking place relatively rapidly in the margins of those primeval seas, thick with possible foodstuffs, almost like the broths we now use to culture bacteria. Once the process of living began, it was probably an accelerating process. New forms would feed on the debris of older ones. There would be widespread and frequent changes in the chemical constituents of the seas, often, probably, destructive to the living forms whose activity produced them. Once life had begun, there were tremendous possibilities for the development of more effective variations. Any successful form would develop in myriads, and amongst these some fresh and more successful variation would soon emerge and spread in its turn. At some point these 'living molecules' must have become enclosed by some form of surface membrane into a self-contained unit, a true 'organism' that represented the precursor of the cell as we know it. Fatty substances (lipids) were quite likely concerned in the formation of such membranes from the beginning but we know almost nothing of the evolutionary sequence by which individualized and highly organized cells developed.

Only when such fully formed individual microorganisms had evolved was there any possibility that a fossil record of living forms could be laid down. Electron microscopic study of ancient sedimentary rocks puts the oldest visible organism, seen in the South African Figtree formation, at 3 billion years old. They appear like primitive bacteria which must have lived in the absence of oxygen by utilizing the organic molecules dissolved in shallow seas. By this time the process of reproduction had been evolved to allow continued growth and yet ensure that the living unit could retain a more or less standard size. Each unit by growth and division now gave rise to descendant units with the same potentialities as itself.

The next great step around 2 billion years ago was the appearance of photosynthetic organisms, still anaerobic but now using the energy of light to produce oxygen from carbon dioxide. The oxygen produced would have been poisonous to these early anaerobic organisms and everything suggests that it was largely transferred to iron compounds,

converting them from the ferrous to the fully oxidized (ferric) form. This gave rise to banded iron formations somewhat before 1·8 billion years ago. The first nucleated organisms with chromosomes (eukaryotes) are dated at about 1·5 billion years and from this period oxygen began to increase in the atmosphere as well as in the iron oxides that gave rise to the great iron ore deposits of Western Australia and elsewhere. Soon after the 1 billion year mark there was sufficient oxygen to produce a layer of ozone at the top of the atmosphere, so shielding the earth's surface from the damaging rays of the ultraviolet. All authorities seem to be agreed that free oxygen was produced almost solely by the photo-synthetic processes of green plants on land and in the sea. Once the lethal ultraviolet had been blocked off the way was open for new complexities of evolution and in the late Precambrian, 800 million years ago, the first many-celled animals (Metazoa) appear in the fossil record. For the remainder of this chapter we will consider the subsequent evolution of these animals and of the processes by which they defend themselves against infection.

*

Our concern is with the great evolutionary discovery that the bodies of other organisms represented a concentrated and easily obtained source of foodstuff for growth. It was from that discovery that the animal kingdom developed. Practically without exception all animals live at the expense of some other organism. The simplest one-celled animals, the protozoa, of which the amoeba is the conventional example, live mainly by taking in and digesting bacteria or microscopic green plants (algae). Every animal from the protozoan upwards must obtain the amino acids it needs to construct its own proteins from the proteins of some other species of organism. Other nutritional requirements vary from one group of animals to another but the protein is what concerns us primarily and, however important they are in other contexts, we can for our present purposes forget about fats, carbohydrates, vitamins and the rest. Protein is the working stuff of life in the sense that all enzymes, all the contractile elements in motile cells and the essential fibrous framework of the tissues are of protein composition.

Without too great an oversimplification we can consider food taken, as bacterium or diatom, into the substance of an amoeba as a fragment of protein. It is protein which is totally different in quality from that of the amoeba itself and if it is to be used it must be broken down by enzymes into its constituent amino acids. This takes place within a small vacuole in the substance of the amoeba's cytoplasm, itself largely composed of protein. The particle of alien protein must be broken down without damaging significantly the animal's own protein. Somehow the amoeba must recognize that the particle taken in is different in quality from its own proteins. Such 'not-self' proteins will be broken down by enzymes which are inactive against 'self' proteins. This introduces the concept of recognition of 'foreignness' which runs like a thread through the whole understanding of immune reactions.

It sometimes happens that the presumptive food particle can resist the enzymes that should dissolve it. More rarely still it may find itself capable of multiplying within the cytoplasm using the constituent molecules for its nutrition. If this goes on the amoeba will die and disintegrate, giving us the most primitive model of an infectious disease. In a real sense every infectious disease of an animal, including man, reduces to the question of 'eat or be eaten'.

We may very briefly look at the implications of this as we come up the evolutionary ladder. In the simplest multicellular animals, the sponges, the mode of nutrition is essentially the same as that seen in the protozoa. Small organisms in the water still constitute the food supply, and these are taken up by cells, themselves very like certain types of protozoa. Within the cells they are digested and the debris discarded much as in the unicellular forms. The chief advance is in the coordination of cells to form an apparatus for the more efficient collection of food. The whole sponge forms a primitive pump, drawing in water with its contained microorganisms through a multitude of pores and discharging it through a small number of larger channels. In the process, the living food particles are taken up by the cells, and waste products are washed out from the tissues. The bulk of the living cells in the sponge are feeding cells, but there are some cells which have other functions to perform, and which receive their nourishment indirectly from the feeding cells. From our angle the significant cells are wandering cells which move about in the jelly-like structureless material at the base of the feeding cells. Their function seems to be partly to transfer nourishment from one part to another, but in addition they have the capacity to engulf and digest any food particles which may enter the deeper substance of the sponge. Such accidental entrants may serve as an additional food supply, but it seems clear that a major function of these wandering cells is to act as protective scavengers and remove any microorganisms which might otherwise utilize the tissue of the sponge as a source of food. This is the first indication in the animal kingdom of a mechanism to deal with invasion of the tissues by microorganisms.

Metchnikoff, who was the first to recognize that the study of comparative physiology might throw light on the nature of infection and immunity, called these wandering cells 'phagocytes', and this name can be generally applied to all such cells in whatever animal they are found. Metchnikoff studied a wide range of vertebrates and invertebrates but his phagocytic theory of immunity was apparently first derived from his observations on the water-flea (*Daphnia*). This is a small transparent crustacean living in fresh water and about the size its name would indicate. It is transparent enough and small enough to be observed directly in the living state under the microscope.

Metchnikoff observed the interaction between the water-flea and the spores of a primitive type of fungus which were numerous in the water of his aquarium. These spores were small, narrow and sharply pointed bodies which, when they were swallowed by the crustacean, were liable to penetrate the wall of the digestive canal and pass into the body cavity. When this occurred, the wandering cells of the body fluid were called into activity. They swarmed around the invading particle, covering it with their living substance and partly engulfing it. Soon it was evident that digestion of the living spicule was taking place. Its smooth shiny surface became pitted, its ends rounded and in time it disintegrated into a walled-in mass of brownish debris.

In the 1960s there was something of a reawakening of interest in the defence reactions of the invertebrates with results that on the whole confirm what Metchnikoff found in the water-flea. In insects there is a body cavity containing fluid and suspended cells (haemocytes) with a primitive contractile heart to maintain a rather inefficient circulation. The haemocytes are mainly concerned with defence against micro-organisms and larger parasites. They have the inevitable capacity to 'recognize foreignness', becoming attached to or taking into their cytoplasm any foreign particles that may enter the body. Many of them are loosely attached to the wall of the cavity, particularly in the neighbourhood of the dorsal heart, and when there is a call for more haemocytes, multiplication takes place in these loose accumulations of cells. There is no substantial evidence in any invertebrate of *specific* immunity, i.e. of recovery from infection with one species of micro-organism being followed by an increased ability to deal specifically with that particular organism. Earthworms however have recently become popular objects of study in these regards and claims have been made that there is some degree of specific immunity mediated by haemocytes in these animals. However, the basic capacity to recognize foreignness and subsequently to react more effectively against previously experienced foreign material really developed only with the most primitive vertebrates, those eel-like blood-sucking creatures, the lampreys and hag-fishes. In all vertebrates from these cyclostomes upwards, we find the same essential capacities that we recognize as 'immunological' in birds and mammals including man.

To anticipate later discussion we may say a little here about the additional faculties that in the vertebrates have been superimposed on the simple haemocyte reactions of insects and the like. Almost in the way that as civilization develops more and more specialized occupations appear, so it seems that, as the vertebrates evolved, cells developed into increasingly specialized types. This holds particularly in regard to immunity. The haemocytes, which appear to be the only mediators of defence in the invertebrates, may well be found on detailed study to comprise two or three distinct types, but the situation is much simpler than in the vertebrates. In mammals like ourselves there are at least eight visibly different cell types involved and amongst these a very wide range of functional differences. An attack of measles calls forth a large population of cells capable of reacting only with measles virus and its products. To produce these cells there is an elaborate system of organs: thymus, spleen, bone marrow and lymph glands, and a blood circulation capable of bringing the cells to bear where they are needed.

Perhaps the most interesting new development in immunology is the recognition that immunity may be something altogether more subtle than we imagined thirty years ago. 'Information' and 'recognition' are two words that seem to be used more and more in biology, and recognition refers not only to the recognition of foreignness that we observe throughout the animal kingdom. For an organ to function efficiently each cell must function in relation not only to the cells which impinge immediately upon it but also to a wide variety of other cells whose activity is relevant to its own. On every cell surface messenger molecules from other cells must impinge and the information they carry be recognized. All the phenomena of defence and immunity must derive from these cell surface functions. It is particularly in the field of malignant disease, of cancer, that these intercellular recognitions break down and one of the most interesting modern fields of immunology is the immune response to cancer cells as they arise in the body. But in this book that is not our business.

This sketch of the evolution of the defence reactions of the animal body against invasion by microorganisms can supply only the roughest outline. In essence, it is little more than a reiteration adapted to a particular range of circumstances of the theme 'eat or be eaten'.

Infectious disease is no more and no less than part of that eternal struggle in which every living organism strives to convert all the available foodstuff in its universe into living organisms of its own species.

Life is everywhere, and every nook and corner that can be exploited for a livelihood is filled. Every living organism is potential food for some other organism, and even if we confine ourselves to the animal kingdom we find an infinite variety of parasitic and semi-parasitic organisms which live at the expense of larger organisms. Every such parasite, be it virus, bacterium, worm or insect, is as much the product of adaptive evolution as its host. The mere fact of its present existence is positive evidence of success in the struggle to survive, and surely indicates a long and detailed adaptation to the particular conditions of its parasitism. The host species of every one of these parasites has also succeeded in surviving. In general terms, where two organisms have developed a host–parasite relationship, the survival of the parasite species is best served, not by destruction of the host, but by the development of a balanced condition in which sufficient of the substance of the host is consumed to allow the parasite's growth and multiplication, but not sufficient to kill the host. For such a balance to develop, long periods of interaction and selection between the two species must have elapsed, and any protective adaptations on the part of the host must often have been countered by adaptive change in the parasite.

The bacteria have been the dominant form of life on earth for some 3 billion years. They began as autonomous organisms capable of survival and propagation in a lifeless environment. Gradually, as higher forms evolved, some of these bacteria adapted to a parasitic existence in which they learnt to feed off other creatures. Just as those higher animals developed mechanisms for defending themselves against bacterial invasion, so in turn have the parasitic bacteria evolved offensive weapons to deal with the defenders. Such qualities are best dealt with in later chapters but a brief mention is needed here of the way the disease-producers have tended to become incapable of living in the simple environments of their ancestral forms and have come to depend on the abundant foodstuffs in their hosts' tissues or secretions. Some of these diminished descendants are capable of growth only within living cells; they are the most obligate of parasites. Amongst them are the rickettsiae, which cause typhus fever and a large group of similar diseases, and the chlamydiae, which is the name now used for the group of agents that includes psittacosis, mentioned in the first chapter. There are many microbiologists who believe that the 'true viruses' of Chapter 5 are also diminished descendants of bacteria. On this view they represent the logical limit of parasitic degeneration, having discarded everything but what was needed to allow reproduction of their nucleic acids. Others look for an evolutionary origin of viruses from genes or organelles of higher cells, that have developed a partial independence of their own.

In this connection it is interesting to consider recent evidence that certain organelles of higher animals and plants may indeed have evolved from bacteria and algae. Evolution is not a one-way street. During the long march of history there may well have been cross-roads at which circumstance favoured moves in either direction. In all animal cells there are small elaborately organized structures, the mitochondria, which carry many of the enzymes that dominate the processes of cell metabolism. These mitochondria grow and divide with a considerable degree of autonomy. Their activities are largely under the control of the cell nucleus but in addition each mitochondrion has its own DNA. This is in the form of a closed circle similar to bacterial DNA. The inference that mitochondria originated as primitive bacteria which aeons ago established a symbiotic relationship with more advanced cells, is now generally accepted. An even more significant event in evolution was the development in similar fashion of the plastids, the chlorophyll-containing granules found in the cells of all green plants. The evidence suggests that green plants arose at various stages in evolution by the incorporation of algae into symbiosis with primitive plant cells. These algae were probably of several types but like the blue-green algae of today all had a primitive 'prokaryotic' form of genetic arrangement closely resembling that of bacteria. The establishment of such symbioses with more developed organisms (eukaryotes) and their subsequent evolution into the green plants provided the main source of the oxygen which progressively accumulated in the atmosphere.

Whether we look at infectious disease primarily as medical microbiologists concerned with understanding and circumventing the microbial aggressors or as immunologists studying the reactions of the host, we shall need above all to apply ecological and evolutionary principles. Throughout the book this will remain the central theme. In the next three chapters we shall introduce the three main groups of disease-producers, bacteria, protozoa and viruses, and then move on to the processes of defence that the body uses against them. There is much more to these than the apparently simple phagocytic response that Metchnikoff described. Immunology undoubtedly grew out of the practical needs of medicine but it has progressively broadened its biological basis. The modern academic approach is concerned on the one hand with what can almost be called the ecology of cellular populations within the body and on the other with the biochemistry and genetics of protein synthesis. Both topics have evolutionary backgrounds of great interest. The medical aspects of immunity have moved away from the centre of scientific interest as in the last ten years immunology has begun to develop its full status as part of fundamental biology. It is almost disappointing to anyone who knew the great days of medical bacteriology and immunology, that the new insights have come almost wholly from areas other than infectious disease.

3. Bacteria

While there are parasitic and semi-parasitic forms amongst all classes of living organisms, those responsible for infectious disease are practically limited to three great groups, the bacteria, protozoa and viruses. We should mention in passing that there are two other groups of parasites, the worms and the fungi, which may cause disease in man of much the same general quality, but few of them are of great importance outside the tropics. Among the medically significant worms for example are *Filaria*, which produces tropical elephantiasis, *Schistosoma,* which is one of the plagues of Africa, and hookworm, which is ubiquitous in tropical and subtropical countries. Then there are a few infections with fungi and yeasts – ringworm of the hair is the best known – which on the whole resemble those of the less actively invasive bacteria. All told, however, these various left-overs do not produce more than a small fraction of the serious illness and death for which bacteria, viruses and protozoa are responsible. In this outline of the general character of infectious disease we will restrict ourselves to the three great groups, beginning with the bacteria.

The general conception of a bacterium is familiar to anyone with a smattering of biological knowledge. They are very small rods or spheres whose size can be measured in μm (1 μm = 1/1000 mm = 1/25000 in). Some of them are surrounded by a fringe of mobile hairs (flagella) by which they can move actively; others have flagella only at one end. A majority have no such appendages, and can move only passively. Inside the body of the bacterium there is little detail to be seen with an ordinary microscope. Electron microscopy however reveals several types of subcellular organelles and a very atypical 'nucleus' without a nuclear membrane. The bacterium is enclosed within a rigid cell wall, largely composed of lipo-polysaccharides, which lies immediately outside of and provides physical support for the cell membrane, the living surface of the organism. By using the enzyme lysozyme the cell wall can be dissolved allowing the bacterium to round up into a ball of protoplasm, a protoplast, which is highly vulnerable to osmotic shock but in suitable environment can still grow, multiply and recreate the cell wall. The structural components of the bacterial wall are specially important in pathogenic bacteria as it is their patterns which form the antigenic determinants that are involved in any immune response to bacterial invasion.

A simple form of sexual union between certain bacteria can occur and genetic material can be transferred from the 'male' to the 'female'. This has allowed some brilliant studies of the genetic control of biochemical processes in selected bacteria, but it seems very unlikely that sexual reproduction plays any major part in the life of bacteria in Nature. For practical purposes we can regard bacterial reproduction as a simple matter of growth and division: the bacterium elongates, a central

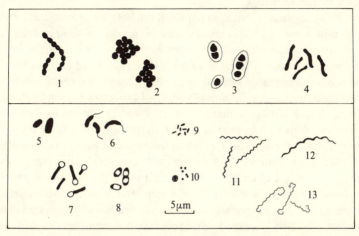

Fig. 2. Outline drawings of bacteria to scale. 1, *Streptococcus*; 2, *Staphylococcus*; 3, *Pneumococcus* with capsule; 4, Tubercle bacillus; 5, *E. coli*; 6, Cholera vibrio with flagellum; 7, Tetanus bacillus with spores; 8, Plague bacillus; 9, Rickettsia of typhus fever; 10, Chlamydia of psittacosis; 11, *Treponema* (syphilis); 12, *Borrelia* (relapsing fever); 13, *Leptospira*.

constriction forms, and what was one is now two. This is the only method of reproduction followed by most of the bacteria that produce disease. A few have an alternative method adapted to allow survival over a period in environments unfavourable for growth. The organism develops a round hard-shelled spore which can survive in dry conditions for years.

Bacteria have quite extraordinary powers of biochemical synthesis. It is easy enough to grow billions of any common bacterium, wash the mass in a centrifuge and then dry the organisms down to a white powder. Nowadays bacteriological chemists think nothing of obtaining several pounds of dried bacteria for analysis of whatever substance they may be interested in. It is only a slight distortion of the facts to say that the dried bacteria contain almost the same components as would be found in a similar mass of dried human tissue, liver or kidney, for example. There is an extraordinary basic similarity in the composition of living matter over the whole range of organisms. The proteins are built up of the same amino acids. The nucleic acids – DNA, messenger RNA, ribosomal RNA and transfer RNA – are basically the same as in higher forms and have in fact been the material on which modern understanding of reproduction and protein synthesis has been based. True fats, phospholipids and sterols are all found, as are most of the vitamins that are needed by human beings.

As living organisms, bacteria require food, both for growth and for the production of energy. This energy may be needed in some species for actual movement, but the more important requirement is to maintain its own living structure. But whereas all higher plants and animals obtain their needs in the matter of food and energy in what are chemically very stereotyped fashions, an immense variety of processes is found amongst bacteria. There are some that obtain their energy by the oxidation of sulphur or of free hydrogen; some can derive their supply of nitrogen from the atmosphere. Most free-living bacteria can grow when supplied with little more than all the necessary elements in some water-soluble form, plus a source of energy, e.g. in a solution of glucose with nitrate or ammonium ions as a source of nitrogen and a simple salt mixture containing the ions phosphate, sulphate, potassium and magnesium with traces of iron and copper. Such bacteria are everywhere, in soil, air, water, and particularly in any type of dead organic material. They fill many important roles in the processes by which the various chemical elements pass into and out of living matter, but their pre-eminent function in nature is to decompose dead organic substance to a form in which it can once again be utilized by plants and indirectly by animals.

The bacteria that cause disease nearly always have more complex requirements; they may for example require several amino acids ready made and some of the vitamins. It is believed that most pathogenic (disease-producing) bacteria are derived from more robust free-living forms which in the course of taking on a parasitic mode of life have lost their ability to synthesize some essential components. These must be supplied ready made. When we talk about these food requirements for different bacteria we are referring to results obtained in biochemical laboratories, where highly purified salts and organic compounds are available. In nature bacteria live in highly complex environments like soil or any collection of dead organic material. Chemically there will almost always be available in such environments all the elements needed for bacterial growth. Often, however, some essential nutrients may be locked away in the proteins or other large molecules of organic material. If it is to be made accessible, appropriate enzymes capable of breaking down the material will need to be produced by bacteria in the environment. In practice all sorts of bacteria accumulate in decaying matter, and in one way or another the original substance is converted into the bacterial bodies which in their turn die and are used as food by other bacteria. In the process there is a steady loss of carbon dioxide and nitrogen to the air and of soluble salts to the soil. The end result when, say, a piece of meat is buried is the accumulation of a little more phosphate and other salts in the neighbouring soil, a small amount of indigestible bacterial polysaccharide and some increase in the complex brown substances that make up a characteristic part of humus.

Bacteria contain no chlorophyll, so that they cannot, like the green plants, utilize the energy of sunlight. They must derive their energy from the chemicals they absorb, often, but by no means invariably, by oxidation of sugars. The biochemical processes concerned are quite complicated, involving a long sequence of steps, each catalysed by a different enzyme. There are other almost equally complex chains of reaction involved in the synthesis of countless cell components, but we have only a partial understanding of where the enzymes concerned are located in the cell and how the 'production lines' are organized. In the present connexion all that need be stressed is that the living bacterium is a nest of enzymes with a turnover of energy that may be more than a hundred times as great as for the same weight of living human tissue.

As far as the physiology of bacteria is relevant to the quality and control of infectious disease in man this brief outline of bacterial metabolism would probably suffice. But we are trying to present infectious disease as part of biology and it would be inappropriate to pass over unmentioned the postwar 'era of enlightenment' in bacterial genetics which may be said to have reached its climax within the last year or two and is now perhaps poised for the emergence of major generalizations.

The essence of bacterial ecology and evolution is the utter expendability of the individual, the clone or the species. In fact it is this expendability which has made it quite impossible to define a bacterial species. A certain pattern of structure and function in a bacterium may be ideal for a given environment but any change in the environment is likely to mean that some slightly different bacterium would now flourish better than the original. If such an organism is available either as a casual 'contaminant' from somewhere else or as a mutant of the originally dominant form, there will be a very rapid change in the population. The new form will dominate the situation, the old one will vanish, comforted perhaps by the thought that in a changing world its successor too is bound to be replaced shortly by something else. The basic requirement for survival is a wide potential for genetic variation with at the same time sufficient genetic stability to allow any advantage to be fully exploited. In a sense we could say the same about any group of organisms, fish or mammals, grasses or conifers. It is after all the basis of the whole Darwinian concept of evolution. But in bacteria the speed of the process is vastly greater. Doubling time is measured in minutes, population numbers in billions per millilitre. In fact bacteria can provide us with a uniquely convenient model for the study of evolution in action. In a few days one can demonstrate changes in a bacterial culture of an order that would need millions of years in man.

The genetic system of bacteria is appropriate to such evolutionary needs. In place of the chromosomes of higher forms, the genetic information in those bacteria that have been studied is carried in a closed ring of DNA. There are now elaborate 'genetic maps' of the bacterium *Escherichia coli* in which the linear order of several hundred genes has been set out on a circular diagram corresponding to the ring of DNA. Indeed there are photographs of the actual DNA obtainable by sophisticated techniques which confirm that physically it is in the form of an endless double thread. DNA replicates itself precisely to pattern most of the time but accidents happen once in every few millions of duplications. Accident or error means mutation. Most mutations are lethal but some changes in the DNA are compatible with survival and reproduction of the organism. According to the nature of the change the organism will have new qualities and very occasionally the difference will provide an advantage in a constantly changing environment. Broadly speaking, genetic errors take one of three forms: point mutation, in which one nucleotide, and hence the gene of which it is a part, is changed; deletion, in which a sequence of nucleotides is lost; or addition (integration) of a segment of DNA from some other source. Integration and deletion necessarily require appropriate enzymes by which the DNA ring can be opened and subsequently reconstituted. By the elaboration and utilization of such mechanisms of inheritable change, an almost unlimited flexibility was provided for bacterial evolution.

Whenever accident or error occurs the new DNA and the changed organism for which it is responsible must face the test of survival. In any flask of turbid bacterial culture there will be among its billions, many thousands of mutant organisms with anomalous DNA being so tested. The vast majority of the anomalous forms are 'unfit' and vanish without trace. But if by some combination of circumstances one of the accidental configurations introduces a new and more effective way of making a living a new and successful competitor will emerge. As is so often the case in biological matters, 'it only needs to happen once'.

The successful novelty need not always result from a change in the main circle of bacterial DNA. Smaller circles of DNA known as plasmids are also present in the bacterium and have some important functions. Many of the queer results that emerge from experiments on bacterial genetics concern interactions of these smaller rings of DNA with the main circle. Plasmids also are subject to mutation. Any change which makes one of the minor rings better able to survive as such can also be ranked as successful.

This may be the place to mention the growing opinion amongst biologists that from this genetic flexibility of bacteria, there emerged two of Nature's most successful inventions, sex and viruses. Like sex in higher forms, plasmids present the organism with an alternative to mutation as a means of acquiring or generating genetic diversity. For instance transfer of a small circle of DNA from another organism can under some circumstances confer resistance to antibiotics which is of great practical importance and is discussed in Chapter 12. The bacterial viruses may well be descended from plasmids which acquired the capacity of easy transfer from one bacterium to another. Their independent evolution gave rise to the 'virulent' bacterial viruses which can invade, multiply within and destroy their bacterial host. There are however 'temperate' viruses, the DNA of which occasionally becomes permanently integrated with the bacterial DNA. Such 'episomes' are to all intents and purposes bacterial genes.

Since bacteria swarm in all dead organic material, it is only to be expected that some species will exploit the organic debris in or on the bodies of living animals. The residues of digested food in the lower parts of the digestive tract provide a particularly suitable source of food for many species, and it is calculated that about half the bulk of faeces consists of the bodies of bacteria. An immensely active struggle for existence amongst probably hundreds of species of bacteria goes on daily in our bowels, usually without interfering in the slightest with our existence. There are other less obvious localities also available in the body, the linings of the mouth, nose and throat, for example. Bacteria from outside are naturally always lodging here, and there are various mechanisms designed to clear them away. Mucus and saliva are secreted, and in part wash the surfaces clear of bacteria. The surface cells of these linings, and particularly those of the mouth, are constantly being renewed and their places taken by others formed beneath them. Thus there is always a certain amount of non-living material available for nutrition in the mouth, nose and throat, and certain bacteria have become adapted to a life here. They live mainly on the dead or nearly dead cells lining these mucous membranes and normally cause no disturbance.

Then we have to consider the scavengers of the skin. Here too the outermost cells are dead and a continuous renewal from beneath is going on. A food supply is available, and automatically groups of bacteria have developed to utilize it. This applies particularly to the more sheltered parts of the skin, where the surface cells are not so readily rubbed off and where they are liable to remain moist with sweat for considerable times, for example, in folds of skin or between the toes.

In these three groups of situations we have bacteria constantly present, not actually growing in living tissues, but in immediate contact with them. It would be curious if at some time or other these normally harmless organisms did not overstep their bounds and try to gain their living at the expense of the adjacent living cells. That seems to be the way in which many of the common disease-producing bacteria have in fact evolved. Amongst the common human diseases caused by bacteria, many are due to organisms obviously related to those found normally in one or other of these situations. And it is particularly impressive that when such bacteria produce disease the region primarily involved is almost always that which normally harbours those harmless bacteria which the pathogenic type resembles.

It is illuminating to consider the various groups of harmless bacteria that lurk about the body, and the related more dangerous forms, in a little more detail. Taking the skin first, we find one predominant type of bacterium, the *Staphylococcus*. Coccus is a general term for any bacterium of spherical or nearly spherical form, while the prefix refers to the habit this particular type has of growing in grape-like clusters. Cultures from any portion of the skin will always grow staphylococci, and it is natural that the common local infections of the skin, pimples and boils, should nearly always be due to the same type of organism. We must not be too precise about this relationship. The staphylococci we isolate from a boil are not quite the same as those we get from healthy skin, nor are the more dangerous staphylococci limited to producing skin infections. Blood-poisoning by these bacteria, though rare, was one of the most uniformly fatal of all blood infections in the days before penicillin. The acute infections of bones which in pre-antibiotic days were common in childhood were also due to infections with staphylococci.

The lining of the nose is essentially an enfolded area of skin and this is another site where staphylococci are nearly always present. Very often the disease-producing type will be found there without causing any inconvenience. In the not uncommon episodes when serious skin infection spreads through a hospital nursery of newborn babies, it is often found that the infection stems from staphylococci carried in the nose of one of the nurses.

In the mucous membranes of the mouth, throat, tonsils, etc., we find a large number of bacterial species. Most of them are of the round (coccus) type. A group in which the cocci are strung together like beads on a necklace is particularly common in the mouth. This is the *Streptococcus* of which there are many varieties, some harmless, some responsible for serious infections such as scarlet fever or acute tonsillitis. Another type of coccus which is found in most human throats is the *Pneumococcus,* so-called originally because it was the commonest cause of pneumonia. We know that it is a common inhabitant of the throat which may invade the lungs when the body's resistance is unduly low. In somewhat similar fashion the *Meningococcus* is not infrequently found in normal throats, but only causes trouble when it spreads into the coverings of the brain and produces meningitis. There are many rather similar, but as far as we know harmless, cocci of this group in the nose and throat. The names *Pneumococcus* and *Meningococcus* remind us that our present point of view neatly reverses that by which the knowledge of these bacteria was gained. Medical bacteriologists have naturally never been particularly interested in bacteria as such. Their interest has been human disease – what caused it, and what could be done about it when the responsible bacterium was isolated. When a particular organism could be isolated from the sputum of 90 per cent of cases of pneumonia, it was natural to call it the *Pneumococcus*. It was only later that bacteriologists began to recognize that the same bacterium was present in the throats of thousands of people without pneumonia.

The spherical or oval form characteristic of the various types of cocci is by no means the only shape to be found amongst bacteria. Probably the commonest of all the forms is the rod-shaped 'bacillus'. In fact, there are so many of these that they have to be divided into more than a dozen different genera. Some of them are present as part of the 'normal bacterial flora' of the mouth and throat but they are much more characteristic of the bowel. The dominant group of intestinal bacilli are anaerobic, i.e. they can grow only when oxygen is excluded from their environment. They are very numerous in the intestine, where oxygen is naturally lacking, but only become dangerous when introduced directly into a wound. Gunshot wounds are particularly liable to infection with anaerobic bacteria, especially when they are contaminated with soil which has been manured with animal faeces. Tetanus and gas gangrene were two very common sequelae to serious wounds received in France during the First World War. Both result from infection with anaerobic bacilli, but both must be regarded as being merely accidental results of the growth of the bacilli in tissues they could never naturally infect. Such infections are not really comparable, either to those we have already mentioned, or to those produced by members of the other great group of intestinal bacilli, which we must now discuss.

When the early bacteriologists made their cultures from faeces, there was one type of bacillus which almost always predominated on their culture plates. This was called *Bacillus coli communis* – the common bacillus of the colon, or large intestine – but is now known as *Escherichia coli,* the '*E. coli*' on which the science of molecular biology was built. Today we realize that this species covers many varieties ('types') of bacteria, and that even so, members of the group represent only a very small proportion of the bacteria in the intestine. Their apparent preponderance is due to the fact that suitable methods have not yet been developed for the cultivation of many of the intestinal anaerobes. In their natural habitat most present-day 'coliforms' do no harm, but there are some types which can cause outbreaks of infantile diarrhoea. The bacteria responsible for the more serious intestinal infections are members of two related genera, *Salmonella* and *Shigella,* which, though they may well have evolved originally from coliforms, have specialized as disease producers for centuries. The bacilli that cause typhoid fever and dysentery are the most important examples in human disease. Similar types have evolved to attack other animals, and not infrequently we hear of human infections caused by bacteria normally infesting rodents or domestic livestock. Several outbreaks of food poisoning have been traced to eating meat from infected animals or food contaminated with bacteria present in the faeces of rats or mice.

These examples should be sufficient to show how bacteria which are normally mere scavengers may develop a capacity to attack the living tissues adjacent to their normal habitat. Once they have developed such powers, they may of course evolve specifically as pathogens, developing clearly demonstrable 'infectiousness' and perhaps changing their preferred site of attack and their requirements for growth. For those we have mentioned so far the predilection for one particular site of attack is clear but this may be lost on further evolution.

Anyone with an ordinary layman's knowledge of medicine will recognize that the various bacterial infections we have been discussing are not equally 'infectious', in the ordinary sense of the term. Diphtheria, scarlet fever and typhoid fever are obviously infectious; boils, wound infections and infections of the urinary passages with *E. coli* are non-infectious in the sense that they are not readily passed to persons in contact with the patient. It is a generalization with only minor exceptions that infectious diseases caused by bacteria are due to species which have become specialized for a parasitic existence and in the process have developed fastidious nutritive requirements. Instead of being able to synthesize all their components from the simplest raw materials some precursors must be obtained ready made from cells or tissue fluids. They are correspondingly more difficult to grow on artificial media than robust synthesizers like the staphylococci and *E. coli*.

Many of these specialized human pathogens, such as those causing diphtheria, tuberculosis, cholera and plague, will be discussed in detail in appropriate places throughout this book. We should take this opportunity however to make brief mention of some other bacterial parasites of man, all of them fastidious in their growth requirements. We can take first the agents responsible for leprosy and syphilis. Both are easily visible under the microscope, the leprosy bacillus as a slender rod closely resembling the tubercle bacillus, the spirochaete of syphilis as a spiral organism. Neither can be cultured on media in the test tube but both can now be grown successfully in laboratory animals. Syphilis was transferred to the rabbit many years ago by inoculation of material from a human venereal sore into the rabbit testis. There large numbers of spirochaetes develop, there is moderate swelling and inflammation and the infection can be transferred readily to other animals by the same type of inoculation. It is interesting that laboratory accidents have shown that even after twenty years' transfer from one rabbit testis to another the spirochaete retains its capacity to produce typical human syphilis. Methods for handling leprosy in the laboratory had to wait until 1967. Then it was found that by inoculating leprous material into the foot-pads of mice the bacteria would slowly multiply. Very recently it was shown that when part of the mouse's immune mechanism was paralysed by removing the thymus at birth the response to leprosy bacillus inoculation was more striking and provided a laboratory model of lepromatous leprosy which is one major form of the human disease.

The next group of organisms to be mentioned are the mycoplasmas. These are minute bacteria which are very fastidious in their growth requirements but can be cultivated on special media. They lack the cell wall of larger bacteria and are correspondingly more susceptible to environmental damage. Nevertheless they are common and in recent years have become rather notorious for the frequency with which they appear in cell cultures during investigations of diseases of unknown cause. The most important disease for which they are known to be responsible is pleuropneumonia of cattle. The only human one which is fully established as a mycoplasma infection is 'atypical' pneumonia, now more usually called mycoplasma pneumonia. The mycoplasmas are the smallest free-living organisms and carry only a few hundred genes. Since they can be found in sewage and compost as well as in association with animal cells it is by no means certain that they have evolved from ordinary bacteria. Indeed one view is that they may represent the surviving descendants of an exceedingly primitive form of life.

Another assemblage of microorganisms which probably have a slightly better claim to be degenerate descendants of bacteria contains two important causes of disease, the rickettsiae and the chlamydiae. The rickettsiae make up an interesting group of pathogenic microorganisms that look very much like small bacteria but can only be grown inside living cells. They are responsible for typhus fever, one of the traditional plagues of war and poverty, plus a wide range of other fevers, some of them trivial, others just as lethal as typhus. As we shall discuss in later chapters, they are primarily parasites of blood-sucking mites, ticks or fleas, but have found a new ecological niche, as producers of human or mammalian disease. One of the first and best of books about infectious disease for the general reader was Zinsser's *Rats, Lice and History* which told the story of typhus fever and the rickettsiae up to 1935. Since then vaccines have been developed and de-lousing with DDT. Together they ensure that armies should never again be plagued by typhus.

There is another group of small 'bacteria' capable of growth only within animal cells which are similar in many respects to the rickettsiae. These are the chlamydiae, which are responsible for three important human diseases. Psittacosis has been described in Chapter 1. The second is trachoma, the eye disease which still afflicts hundreds of millions of people and is by far the most important cause of blindness in the hot dry regions of the world. Lymphogranuloma venerum ranks behind syphilis and gonorrhoea as the third of the major venereal diseases, being especially prevalent in the tropics. Both the chlamydiae and the rickettsiae are, fortunately, susceptible to 'broad-spectrum' antibiotics. This fact together with recent studies on the chemical composition and metabolism of both groups of organisms place them much closer to the bacteria than to the viruses, which will be discussed in a separate chapter.

4. Protozoa

In temperate climates the great majority of infectious diseases are produced by bacteria or viruses; infections by protozoa are relatively very rare. In tropical and subtropical regions, however, we find four human diseases of first-rate importance which are caused by these minute unicellular animal parasites. Malaria is the most widespread and important, and as well we have sleeping sickness, kala azar and amoebic dysentery. There are also many economically important diseases of domestic animals which are due to protozoal infections. Research on the protozoa has followed rather different lines from those of orthodox bacteriological research. Most students of the subject have been trained as zoologists, and a large proportion of their work has been concerned with the microscopic appearance of the parasites and with their life histories in infected animals or human beings. Much less work has been done on the nature and extent of immunity which follows infection than in the field of bacteriology proper.

Protozoa are the smallest animal organisms. They are unicellular, and although they are on the average considerably larger than bacteria, none of the forms which produce disease can be seen with the naked eye. There are vastly greater numbers of completely harmless types of protozoa, and if we include the whole group we find a considerable range of size. The smallest forms of such types as the parasite of malaria are no larger than an average bacterium, while amongst the non-parasitic types there are a number which can be easily seen with a small hand-lens, and are actually a good deal larger than some of the smallest insects. A typical protozoon is easily recognized as a true animal organism, not to be confused with unicellular plants and bacteria. It is a single cell with a well-developed nucleus, usually capable of active movement and feeding by taking into its substance and digesting smaller microorganisms of various sorts. In the simpler forms such as the amoeba, movement and the taking in of food are both functions of the general cell substance. A portion of the semi-fluid protoplasm flows out into a sort of temporary organ, a pseudopodium, which then serves either to drag the amoeba along or to help engulf the microorganisms on which it feeds. When the amoeba becomes inactive, the pseudopodia are withdrawn and the protozoon becomes a featureless spherical blob of jelly. More specialized forms have developed quite elaborate permanent mechanisms to facilitate rapid movement and the more effective capture of food. Some free-living protozoal species can, however, obtain their nourishment wholly from soluble materials in their environment.

Protozoa are found in practically every situation which can provide both water and a food supply of smaller microorganisms. Any collection of fresh water in contact with decaying organic matter will usually contain protozoa of various types. Every amateur microscopist is familiar with *Amoeba, Paramecium* and *Vorticella* from such situations, but there are nearly always many other forms not so easily identified. It is natural that protozoa should be constantly swallowed by all types of animals, and since the contents of the intestinal canal contain both water and an abundance of bacteria, we should expect certain forms to exploit this situation. There is, however, the important difference that life in the intestine must go on in the absence of oxygen. Only those forms which are capable of anaerobic existence can become intestinal parasites. Actually there is probably no species of vertebrate which is free from infestation by protozoa in its intestinal tract. At least a dozen species are known to be capable of living in the human intestine. Insects too are very frequently found to harbour parasitic protozoa in their digestive tracts. Originally such protozoa probably lived on bacteria and particles of semi-digested food, but like bacteria in similar situations, some of them have developed the capacity to live directly on the cells and fluids of their hosts. Our knowledge of the evolution of parasitic protozoa must necessarily be obtained only indirectly by comparison of the habits of the pathogenic forms with structurally similar free-living and semi-parasitic forms. Such studies do, however, provide a substantial basis for deduction, and have given rise to theories which offer an attractive interpretation of what otherwise would be a series of unconnected and very complicated phenomena.

The four important protozoal diseases of man (or for that matter practically all protozoal infections of vertebrates) can be divided into two groups: those in which the infection is transmitted directly from man to man, with the one human example of amoebic dysentery, and those in which the parasite is transferred by an insect intermediary – malaria, sleeping sickness and kala azar. It is generally believed that the first group is derived from protozoa which were originally harmless scavengers in the vertebrate intestine, while the forms in the second group are more or less highly specialized descendants of protozoa playing a similar role in the intestine of insects.

Amoebic dysentery is due to a relatively unspecialized parasite, an amoeba very similar to forms which normally make a living amongst the bacteria and debris of the intestinal contents, and not greatly different from the free-living amoeba of pond water. Instead of being content with a diet of bacteria, the disease-producing amoeba has developed a capacity to invade the tissues of the intestinal wall and draw its nourishment from red blood cells and fragments of damaged tissue. In a serious infection with the virulent form of *Entamoeba histolytica* the amoeba excretes enzymes which break down neighbouring cells – hence the name *histolytica* – and gains its nourishment primarily by absorbing soluble material from its environment. This destructive action results in the formation of ulcers in the wall of the intestine and the appearance of the symptoms of dysentery. From the ulcers some amoebae may enter the blood and be carried to the liver. Multiplication of the protozoa in the liver produces the commonest complication of amoebic dysentery – a liver abscess.

From the biological point of view, we are dealing here with a development very similar to the change from the harmless skin staphylococci to the virulent ones which cause boils and carbuncles. Like the staphylococci, the dysentery amoebae may show wide differences in virulence. Many people may harbour the parasite without suffering any symptoms at all. In these individuals the amoeba probably causes only trivial ulcerations of the bowel. The reasons for these differences are still not understood. The possibilities are, first, that there are inheritable differences in the infecting strains of *Entamoeba* and the severity of a man's attack depends simply on whether he is infected with a highly virulent strain or not. The second possibility is that resistance will vary according to the state of nutrition and general health of the person infected plus, as in all such situations, the effect of genetic or constitutional differences in resistance. A third factor that may well be imporant is the coexistence of bacterial or viral infection in the bowel. It may be that the amoebae find it much easier to initiate infection in an intestinal lining damaged by infection of some other type. Where amoebic dysentery is rife we naturally find many opportunities for all forms of intestinal infection. There is probably a continual entry of new strains of amoebae each, as it were, being subjected to the test of survival in this particular intestine. Sooner or later in one set of circumstances or another the strain with or without the assistance of some other type of infection will find it possible to invade and damage the body. This is one of the infections in which there is little evidence of immunity following an attack, a fact which in itself suggests that the 'standard' amoeba requires some additional factor to allow it to produce serious illness.

We do not know for how long amoebic dysentery has afflicted human beings. It is probable that it was derived from an infection of the common Asian monkeys. The ubiquitous rhesus monkey of India usually carries an amoeba indistinguishable from the human *Entamoeba histolytica* but suffers no ill effects. South American monkeys on the other hand are liable to contract dysentery similar to the human disease when they are artificially infected. It is another example of the rule that long association of host and parasite is evidenced by little or no power to produce serious symptoms or death.

It is not so simple a story when we try to understand the evolution of such specialized parasites as those which cause malaria and sleeping

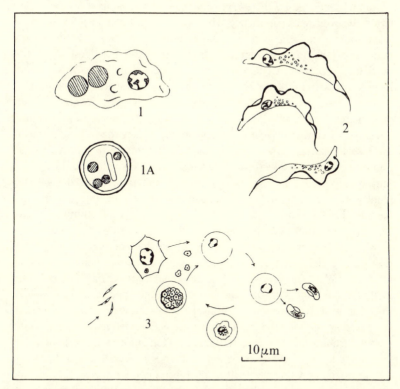

Fig. 3. The three major types of protozoa pathogenic for man. 1, *Amoeba histolytica* (amoebic dysentery); 1A, *Amoeba histolytica* (encysted); 2, *Trypanosoma brucei* (sleeping sickness); 3, The human stages of *Plasmodium* of malaria.

sickness. In the case of sleeping sickness we have an infection, first of the blood, then of the brain, by a protozoon parasite which is transmitted by the bite of the tsetse fly. The parasites are members of a group of actively moving protozoa which in their typical form are called trypanosomes. Many, however, can take on a variety of appearances including a non-motile form analogous to the *Leishmania* which is responsible for kala azar. Current opinion is that there is only one rather highly variable species, *Trypanosoma brucei*, that is capable of infecting man. This is sometimes found in animals but it is clear that the important reservoir of infection is man himself.

There are millions of square miles in tropical Africa where trypanosomes of a number of types carried by tsetse flies make it impossible to keep cattle or other domestic ungulates. Though infection of several types of small antelope may also be fatal, most wild mammals are harmlessly infected by these trypanosomes and serve as a reservoir. The tsetse fly cycle will commence when the insect sucks in infected blood from some vertebrate carrier. Multiplication can take place in several regions of the insect's digestive tract. Everything suggests that this particular group of African trypanosomes evolved from species which lived harmlessly in the far end of the intestine. There are still some which multiply only in this situation, but the important forms that are responsible for sleeping sickness in man and fatal disease of horses have developed a complex cycle which eventually lodges the trypanosomes in the insect's salivary glands ready for injection into a new host. It takes about two to four weeks for this development to take place in the fly, and during this period it cannot transmit the infection. Once the salivary glands are infected, however, trypanosomes will be injected at each feeding. The tsetse fly remains capable of spreading infection for at least three months after the trypanosomes first reach the salivary glands.

We can see some of the stages by which this complex life history of the sleeping sickness trypanosome probably came into being by looking at the simpler life histories of some related protozoa. There is, for instance, one which infects the blood of some species of lizards. The infection is derived from a small biting fly by the very simple process of the lizard swallowing the fly. The parasite is liberated in the intestine and passes through the wall into the lizard's blood, where it multiplies. The next stage is for a new fly to be infected while it sucks the lizard's blood. The protozoa now find themselves, as it were, in their ancestral home and multiply in the lower end of the fly's intestine, behaving just like any protozoal parasites of insects which have no part of their life cycle in vertebrates.

The next step can be seen in one of the African tsetse flies which carries a trypanosome that infects crocodiles. In the insect the protozoon is limited to the lower end of the bowel and the crocodile is infected by contamination of the bitten points with the fly's faeces.

A rather similar, but probably independent, line of evolution was concerned in the development of the South American form of trypanosome disease (Chagas' disease). This protozoon is found naturally in armadillos and other native mammals. It is taken in by a species of bug which has catholic tastes and feeds on men as well as armadillos. The trypanosomes develop in the bug's intestine and pass out with the faeces. They do not get into the insect's saliva, and are not injected when the bug bites. Human infections develop only if the bug is crushed while it is biting, or if scratching of the spot works fragments of faeces into the puncture wound. Chagas' disease in one form or another may involve as many as seven million people in South America. It is variable in its manifestations which include skin lesions and lymph gland enlargement, but the most important clinical effect is chronic and often severe infection of the heart muscle. Diagnosis is often difficult and there is still no satisfactory treatment of chronic infection. It has been suggested by a very distinguished protozoologist that Chagas' disease was responsible for Charles Darwin's chronic illness which has puzzled all his biographers. In the course of his travels in South America during the voyage of the *Beagle,* Darwin records that on at least one occasion he was extensively bitten by the huge *Triatoma* which carries the infection. His illness dated from the time of his return to England and perhaps helped to provide the almost unique combination of circumstances that allowed the idea of evolution to come to its full expression in *The Origin of Species.*

Kala azar, which in the period 1890–1905 was one of the major plagues of South-East Asia, is due to a protozoon that has some resemblance to the trypanosomes but is placed in a distinct group, *Leishmania.* These are responsible for two distinct conditions, localized skin lesions (Oriental sore, Delhi boil) and kala azar in which the parasites multiply in liver and spleen. The species responsible are obviously very closely related; both are carried by sandflies and both may on occasion have an animal reservoir in dogs or rodents. The usual reservoir for kala azar, particularly in India, is man himself. Both the generalized and cutaneous forms are found in Southern Asia, the Eastern Mediterranean and Central America, while the generalized form also occurs in East Africa. The protozoa multiply in the stomach of the sandfly to produce large numbers of motile forms and, much as happens in the flea infected with plague bacilli, some have to be regurgitated before the sandfly can take a new blood meal. Both dogs and human beings can be infected and, depending on the age of the victim and individual differences between protozoan strains, the result may be a limited skin infection (Oriental sore) or a general infection of variable severity.

Malaria needs a chapter to itself and here we need say only enough to indicate its general character and to note current opinion on how the malarial parasites evolved. The organisms responsible for the three main forms of human malaria are all spread by mosquitoes of the genus *Anopheles*. In simplified outline the mosquito injects a tiny malarial form into the blood which carries it to the liver. There it infects liver cells and multiplies to produce a store of descendants from which the red cells of the blood are infected. Multiplication for a few generations in the blood is associated with destruction of cells, each wave of destruction being the signal for a paroxysm of fever. Subsequently sexual forms appear which, when they are taken into a mosquito, unite to give rise to cysts from which a new generation of parasites can reach the insect's salivary gland. At all stages the essential part of the process is multiplication inside a cell to produce a brood of descendants which then break out from the cell and pass to whatever type of cell is appropriate for the next stage in the cycle.

It is highly probable that the malarial parasites are related to the protozoa responsible for the diseases of rabbits and fowls that we call coccidiosis. In these the disease is primarily in the wall of the intestine. The whole process is broadly similar, but instead of infecting a mosquito the sexual form develops into a resistant cyst which passes out with the faeces. Within these oocysts further development and multiplication of the parasite takes place and they then become the vehicle by which infection is passed to another individual of the host species.

In this, as in the other examples of disease due to protozoa, we can only make speculative suggestions about the evolution of the micro-organisms and of the diseases they produce. We shall have more to say about the general problems of infectious disease in later chapters, but here already we can see some outlines of the pattern that will emerge. It is very evident that when a parasite and its host have lived together for very many generations the association is a balanced one with little evidence of damage to the host. The African trypanosomes do no visible harm to the game animals that are their natural hosts, nor the *Entamoeba histolytica* to the rhesus monkey.

Partly because of the absence of simple technical methods for their demonstration, there has been relatively little work on antibodies and immunity against protozoal disease. There is evidence of immunity, but in most instances it is an 'immunity of infection' that disappears soon after the parasites are finally eliminated from the body. Individuals born and growing up in a heavily malarious environment will, if they survive, show no symptoms of malaria in adult life. They are chronically infected, but they have developed sufficient immunity to keep the multiplication of the parasite down to a harmless level. Such an individual who has been free of symptoms for years may still be capable of infecting a susceptible person by blood given for transfusion. It is possible that immune reactions are more important in malaria than was once thought. People in areas of endemic malaria show almost consistently high immuno-globulin levels probably largely due to antibody against malarial parasites or their products. Serum from persons with high immunoglobulin levels has been used with some apparent success in the treatment of cerebral malaria but no serious attempt to prevent malarial infection by any form of immunization has been made. Similar negative or equivocal findings hold in regard to other protozoal infections and control of all these diseases has had to be sought on other lines.

The first is to modify the ecological situation so as to block some point in the cycle by which the parasite survives. The second is to find and use drugs which can damage the parasite while doing minimal harm to the cells of the host. Both approaches will be more appropriately dealt with in later pages.

5. Viruses

The golden age of bacteriology was in the 1880s, when Pasteur had laid the foundation of the science and the solid genius of Koch was providing the technical methods and the scientific point of view. Young men from Koch's laboratory had all the world of infectious disease before them to conquer. They investigated every disease which might be due to infection by some microorganism, and almost invariably found bacteria. The more fortunate workers found the right bacteria, and so established the cause of anthrax, diphtheria, typhoid fever and so on, but others were less fortunate. The bacteria that they isolated from cases of measles, influenza or smallpox turned out to have nothing to do with the disease, and it gradually became evident that there were some infectious diseases which were caused by agents smaller than bacteria and protozoa.

Quite early in the history of bacteriology, a Dutchman, Beijerinck, found that the agent responsible for the mosaic disease of the tobacco plant could pass through the pores of a porcelain filter candle which could keep back all bacteria. This was the first discovery of a 'filterable' virus. Then, just at the turn of the century, two German workers found that the germ of foot-and-mouth disease of cattle could also pass through such filters, and established the existence of filterable viruses causing disease in animals. Then there came a period when any manifestly infectious disease not associated with visible or cultivable micro-organisms was ascribed to viruses. It took a long time before the modern concept of a virus was reached. We now realize that viruses are ubiquitous and by no means always produce obvious signs of disease. There are viruses of mammals and birds, fish and amphibians, insects, higher plants, algae, fungi and even bacteria.

Around 1930 new technical methods rather suddenly made it possible to isolate and characterize a considerable proportion of the viruses responsible for serious human disease. Since then progress has been rapid and when it was realized that the very simplicity of viruses had enormous virtues for biochemical attack on some of the most fundamental aspects of life, virology became literally one of the major branches of biology. It is not too much to say that many of the most significant developments in molecular biology centred on the study of the bacterial viruses active against standard strains of *E. coli*. Another revolution of almost equal importance for the study of the viruses concerned with human and animal infections was the development and refinement of methods for growing mammalian cells in test tubes so that they could serve as hosts within which viruses could multiply. Today virtually all viruses responsible for human disease can be grown, assayed and studied in various ways in such cultured cells – 'cell cultures' or 'tissue cultures'. Activity in virological research in 1971 has moved rather far away from what it was in the beginnings of the modern phase in the period 1930–50. Interest is now very largely directed toward chemical and genetic aspects. The significance of viruses for disease in man and the ecological features of their survival in nature with which we are primarily concerned in this book are currently somewhat out of fashion.

The new approaches have changed the whole outlook on viruses. Even the definition of a virus nowadays is quite different from the old one, which stated that a virus was a microorganism capable of multiplying only within living cells. The present approach is to define a virus as a structure composed of a protein coat surrounding a nucleic acid molecule, either RNA or DNA, which is capable of replication only within living cells. A virus is not an organism; it has no metabolism and is wholly dependent for its reproduction on mechanisms provided by its host cell.

The modern classification of viruses is based on the physical and chemical characteristics of the virus particles, the 'virions', by which infection is transferred from one cell to the next. The basic criterion is the nature of the nucleic acid, DNA or RNA, single- or double-stranded, and the number of genes it contains. Also important is the way in which the protein molecules are arranged around the nucleic acid. In many viruses the protein molecules occur in small aggregates, known as capsomers, which under the electron microscope can be seen to be arranged in the form of a regular icosahedron made up of twenty surface facets, each an equilateral triangle. Other viruses possess a basically spiral structure and are said to show helical symmetry. Some groups of viruses with either type of symmetry acquire an outer membranous envelope of cellular origin as they are being liberated from the cells they infect. Projecting from this envelope are numerous short spikes of glycoprotein that enable the virus to adhere to and infect susceptible cells.

Only a limited number of diseases due to viruses will be discussed in this book and an abbreviated table can be given to show how each of those we shall mention fits into the modern classification.

Table 1. *The commoner viruses of man*

Viruses	Group	Nucleic acid	Diameter (nm)	Shape	Symmetry	Envelope
Smallpox, vaccinia	Poxvirus	DNA	100×200	Brick shape	Nil	+
Herpes, chickenpox, cytomegalovirus, EBV }	Herpesvirus	DNA	150	Spherical	Icosahedral	+
Adenovirus	Adenovirus	DNA	75	Spherical	Icosahedral	–
Warts	Papovavirus	DNA	50	Spherical	Icosahedral	–
Mumps, measles, R.S.V., parainfluenza }	Paramyxovirus	RNA	200	Spherical	Helical	+
Influenza	Myxovirus	RNA	100	Spherical	Helical	+
Rabies	Rhabdovirus	RNA	75×180	Bullet shape	Helical	+
Yellow fever, dengue	Arbovirus	RNA	40	Spherical	Icosahedral	+
Polio, echo, coxsackie, common cold }	Picornavirus	RNA	25	Spherical	Icosahedral	–

Smallpox

Rabies

Measles

Chickenpox

Influenza

Coronavirus

Yellow fever

Bacterial virus T2

Adenovirus Reovirus

Warts

Poliovirus

Parvovirus

Tobacco mosaic virus

100 nm

Fig. 4. Scale diagrams showing the size and structure of important human viruses with typical plant and bacterial viruses for comparison. (From Fenner & White, *Medical Virology.*)

Much of the detail in Table 1 can be presented without attempting to elaborate the significance of the various differences. Taken along with Fig. 4 the table will give an impression of the wide physical differences amongst the viruses responsible for human disease. The size range can best be realized by saying that it covers the gap between a large protein molecule, like haemocyanin, and the smallest free-living microorganism,

55

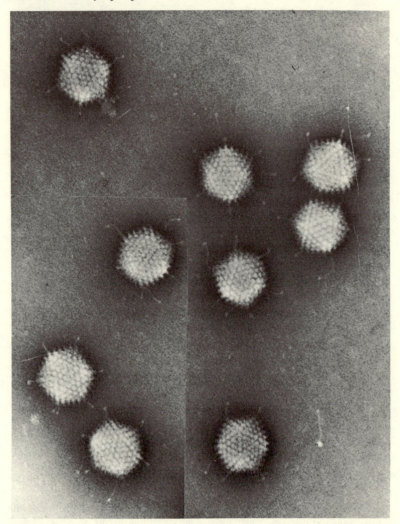

Fig. 5. An adenovirus showing the classical icosahedral form and the fine fibres protruding from each vertex of the polyhedron. (Electron micrograph by courtesy of Dr N. G. Wrigley.)

Mycoplasma. Only the very largest viruses can be seen under the ordinary light microscope. However, under the electron microscope, which has a resolving power of less than 1 nm (10 Ångströms), the fine detail of viral substructure can nowadays be clearly visualized as is evident from Fig. 5.

Our theme of the natural history of infectious disease and the agents, including viruses, that are responsible is not however deeply concerned with the physical structure of viruses nor with how they replicate themselves within the host cell. It is as well to remember that yellow fever was eliminated from Cuba and the Panama Canal Zone before it was known whether a virus or a bacterium was responsible for the disease. Admittedly that was an exceptional situation.

In general the epidemiologist or ecologist interested in a given disease will require a reliable means of deciding whether or not the pathogenic agent is present in any material such as blood or faeces that he submits to the laboratory plus an assurance as to the specific character of the agent. It is almost equally important that one or more ways of recognizing specific antibody in serum should be available. If we restrict our account of laboratory work to these requirements it can be kept at a relatively non-technical level.

To demonstrate the presence of virus in any material one must find a system of cells in which it will multiply with the production of visible changes. One can almost trace the history of virology in the sequence of the methods by which these requirements were fulfilled. Medical virology grew out of bacteriology and the early objective was to 'isolate' the agent of disease and establish its authenticity by applying the tests incorporated in 'Koch's postulates'. Koch, who was the greatest of the early bacteriologists, laid down the requirements for proving that a particular bacterium is responsible for a particular disease: first, that the bacterium should be present in all cases of the disease, secondly, that it should be grown in pure culture, thirdly, that organisms from such cultures must produce the disease when inoculated into experimental animals, from which, lastly the organism must again be recoverable. With the recognition of viruses – which cannot be cultured in Koch's sense – as agents of disease, proof of isolation hinged on the investigator's ability to produce the disease or its equivalent in experimental animals and show that it was transmissible to successive series of susceptible animals.

Forty years ago all experimental work on viruses was based on the use of small laboratory animals which when appropriately inoculated would show symptoms as a result of the multiplication of virus in susceptible tissues. Poliomyelitis in children was obviously a disease of the spinal cord and the only method for its laboratory study was to inject into the brain or spinal cord of monkeys. If the monkey showed paralysis with typical pathological changes in the spinal cord we could be certain that there had been polio virus in what was inoculated. Rather similarly, influenza virus was first studied in ferrets, which are the only animals known to respond to infection with the virus with symptoms and tissue changes in the respiratory passages almost the same as those in the human disease.

Monkeys and ferrets are rather too large and too expensive for convenient use in the laboratory and for a very important phase of virus research the main aim was to adapt viruses so that they would multiply and produce lesions in the most convenient of all experimental animals, the mouse. There are no virus diseases of man that are 'infectious' in the ordinary sense for mice, but there are many which can multiply in one or other tissue if they are inoculated artificially. Most viruses that produce infections of the nervous system can be persuaded to multiply in the mouse brain if they are inoculated there. Influenza virus will multiply in the mouse lung and after a few transfers from one mouse to another will produce pneumonia in the form of red airless regions in the lung. Once a suitable mutant capable of producing easily recognized effects such as death, paralysis or pneumonia had been selected in this way, virologists were in a position to use the animal as a convenient model for studying various aspects of the disease.

The next development was the utilization of chick embryos for virus research. This has some important advantages over mouse inoculation, since there are no bacteria in a chick embryo inside its shell and we can see the effects of virus infection without being worried by doubts as to whether contaminating bacteria were partly responsible. On the whole, embryonic cells are highly susceptible to attack by viruses and there are many different ways of inoculating eggs to allow virus to reach the most suitable of the various membranes and even embryo organs such as brain and lungs. In the early 1930s several viruses were shown to be capable, either directly or after the selection of an appropriate mutant, of producing discrete pocks on the outermost membrane, the chorioallantois. The viruses included smallpox, herpes simplex and influenza and for a period the method provided the most convenient method of assay. A few years later it was found that influenza virus could be isolated by inoculating throat washings from patients into the inner (amniotic) cavity and that for many purposes the outer (allantoic) water jacket was an even more suitable site for influenza multiplication. A little later the yolk sac was also found to have special virtues for the growth of some viruses and particularly for the rickettsiae, which cause typhus fever and similar diseases.

The growth of viruses in mammalian cells maintained in what was then called tissue culture began as early as 1926 but it was not until 1948 that cell culture emerged in its modern form. Embryonic cells and a wide

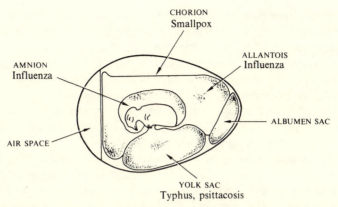

Fig. 6. A conventionalized diagram of a chick embryo about half-way through incubation, to show the membranes used for growth of viruses, rickettsiae and chlamydiae.

variety of adult cells can easily be grown in tubes or bottles if they are provided with the appropriate mixture of salts and nutrients and the necessary precautions taken to protect them from bacterial contamination. The commonest method is to let the cells grow on the surface of a glass container, but there are also ways by which some cells can grow and multiply when freely suspended in the nutrient fluid. If we start from a normal tissue, for example monkey kidney, the usual procedure is to obtain a thin suspension of cells from the kidney by gentle treatment with the enzyme trypsin. These are then allowed to settle to the bottom of a thin layer of nutrient fluid where each cell sticks to the glass and multiplies, producing two descendants (binary fission) about once in every twenty-four hours. In a few days' time when there is a continuous sheet of cells, the virus is added, and in another three to four days it has multiplied and spread to involve every cell in the culture.

There are some cells, especially those which have developed the heightened capacity for multiplication that is typical of cancer cells, which can go on dividing indefinitely in culture. Several such 'lines' of cells are widely used in virus research, the best known being the strain HeLa from a human cancer. The use of such strains has freed virus research completely from its dependence on experimental animals. A constant supply of growing cells is maintained and whenever an experiment is to be set up the required number of tubes or bottles are seeded with the cells and used after an appropriate period of preliminary incubation.

A method of growing viruses is of no particular use unless we can find visible evidence that growth has in fact occurred. Fortunately most viruses kill the cultured cells in which they multiply. Visible signs of cell damage, spoken of as cytopathic effects (CPE), enable the virologist to recognize the growth of virus by simple inspection under a low power microscope. Furthermore, the number of infectious virions in a suspension can be measured by assay methods such as the following. Cells are distributed over the bottom of shallow glass dishes and grown long enough to give a confluent sheet. The virus is added and left about an hour to allow the virions to attach to cells. The fluid is now removed and the cell monolayer gently overlaid with a soft jelly and incubated for a few more days. From each cell initially infected the first generation of descendent particles spreads contiguously to adjacent cells and from them to others further out until after four or five days we have a circular 'plaque' visible to the naked eye. The number of plaques is, of course, a measure of the number of infectious virions that was present in the original material. Such methods can be adapted to test whether a given antiserum neutralizes the virus.

Probably the most important result of the development and routine use of cell culture methods was the sudden realization that relatively harmless viruses were almost as ubiquitous as harmless bacteria. Undoubtedly we should have known better, but it is a fact that until the ECHO (enteric cytopathogenic human orphan) viruses began to turn up in the 1950s the conscious and unconscious assumption of all virologists was that to find a virus you must start with an infectious disease. So when people were testing faeces for polioviruses and found in healthy children a cytopathogenic virus distinct from poliovirus, they called it an orphan, i.e. a virus in search of a parent disease. That was just the beginning. Cell culture methods made it clear that there were hundreds of distinguishable virus types (mainly echoviruses, coxsackieviruses and adenoviruses) which could be found in faeces or in respiratory secretions, occasionally associated with diarrhoeas or colds but more often producing no sign at all of illness. A little later electron microscopists began to be familiar with the appearance of virus particles in thin sections of tissue and the sort of changes that virus multiplication might make to the architecture of a cell. If we look carefully at a large enough number of electron micrographs of sections from any animal, healthy or diseased, we are liable to find a few virus particles. Viruses are ubiquitous, and more often than not are doing the host no harm.

All this fits in well with what had been known for a long time about subclinical infections by viruses and bacteria which were fully capable of producing disease in other individuals. The unexpected orphans discovered by cell culture methods differed only in being normally harmless and producing recognizable illness rarely or not at all. In many ways we can extend the finding from bacteria that the disease-producing forms tend to be related to and perhaps derived from harmless viruses living at the expense of the same sorts of cells. The situation is of course more complex than with the bacteria. No matter how harmless a virus seems, if it is readily transferable from one individual to the next it must be multiplying in and being liberated from the host cells. With some the cell shows no evidence of serious damage; with most a few cells are damaged but no symptoms result. We doubt if anyone has more than a superficial understanding of what determines whether a virus infection, recognizable by the fact that virus is liberated or the individual develops a specific antibody response, does or does not produce symptoms. Most viruses that completely destroy cultured cells are nevertheless restrained in the body by a variety of controlling factors we still know relatively little about. Some of them will be discussed in the next chapter.

We must be clear that not all animal viruses produce visible damage in any culture of mammalian cells. Even with a well-known virus such as rubella only a few cell types allow its multiplication and in fewer still is there damage to the cells sufficient to be recognizable by the usual criteria. Indeed there are other viruses as well as rubella – the common cold viruses for instance – which showed no cytopathic effects in the cultures in which they were first isolated. The fact that multiplication of virus had occurred was recognized only by the use of certain ingenious technical tricks. Even in 1971 microbiologists have failed to grow at least two viruses which can produce fatal disease in man, hepatitis and kuru, and are still open-minded about the possibility that viruses may yet prove to be responsible for a fairly considerable number of obscure human diseases of unknown etiology.

Before passing on to other aspects of virus disease in man and animals we should briefly mention the susceptibility of other types of organism to viral infection. It is true to say that virus diseases are known of almost every living species of organism that microbiologists have had reason to study in detail. Much human virus disease is derived from a reservoir of mammals or birds and is conveyed by an arthropod vector, such as mosquito or tick. When this happens it necessarily follows that the viruses concerned must be versatile enough to grow in insect cells as well as in a considerable range of cells from mammal or bird as well as human cells. Most common insects and other arthropods have their own virus infections, limited usually to a small group of similar species. More interestingly, some plant-sucking insects can be both vectors and hosts of viruses which produce either frank disease in plants or harmless infections analogous to subclinical infection in man. The list could be extended indefinitely, but for our present purposes only two groups other than the human and mammalian viruses need be mentioned. The plant virus, tobacco mosaic (TMV) and the bacterial viruses (bacteriophages) played too great a part in the history of virology to be passed over without comment.

The virus of tobacco mosaic disease, as we have mentioned, was the first virus to be proved filterable, and, ever since, most of the fundamental work on the nature of plant viruses has been done with this example. In the tobacco plant the virus multiplies prolifically, so much so that in the squeezed-out sap of an infected plant the virus constitutes about 80 per cent of the total protein present. In the infected tissues of animals the actual mass of virus present is extremely small, and it is an expensive and tedious process to obtain any significant amount of purified virus for chemical examination. There is no such limitation with tobacco mosaic virus, and since the days in 1935 when Stanley purified the virus in the form of what he called a crystalline protein it has been a favourite object of study. The excitement that greeted Stanley's discovery has faded with the years. We now know the minute amount of RNA inside the virion to be much more important than the protein coat. TMV seen by the electron microscope appears as regular cylindrical rods. In fine structure there is a thread of RNA spiralling down through the substance of the rod which itself is composed of hundreds of uniform protein molecules packed in a regular flat spiral. The individual molecules are rather small with a molecular weight of 18,000 and the complete sequence of amino acids in this protein is now well known.

A spectacular discovery in 1956 by Gierer and Schramm showed that the RNA could be separated from the protein by a simple chemical method and was still infectious. This had great influence on the whole of biology because it showed by direct experiment what was gradually becoming clear for other reasons, viz. that all the 'information' needed to allow virus development, including synthesis of its characteristic protein, was carried in nucleic acid. It was not long before it was found that separation of infectious RNA was also possible in the small animal viruses.

The bacterial viruses (bacteriophages) have been even more significant for the development of molecular biology and for an understanding of the biological nature of viruses. Their importance for infectious disease is only indirect. It is one of the minor ironies of the history of medical science that their discoverer, d'Herelle, thought at one time that it was only a matter of collecting and 'training' suitable bacteriophages to eliminate bacterial disease from mankind. There are very numerous bacteriophages which in the test tube will destroy one or more of the species of bacteria responsible for intestinal disease – the cholera vibrios and the dysentery bacilli, as well as the salmonellas responsible for typhoid fever and some sorts of food poisoning. Nearly half a century ago, shortly after d'Herelle's momentous discovery, there were claims that phage treatment was beneficial and counter claims that it was useless. Probably no carefully controlled trial, by people with a modern understanding of all the conditions in the bowel relevant to a three-sided interaction of bacillus, bacteriophage and the patient's defences, has ever been made. All these bacterial diseases of the bowel can now be treated effectively in other ways and there is no longer any incentive to reopen the possibilities of bacteriophage treatment. Unless something very unexpected occurs d'Herelle's ideas will never be properly tested.

The contemporary relevance of phages to infectious disease depends on their specificity in relation to the bacteria they attack. Following an outbreak of typhoid fever for instance it is often very important to know whether a group of half a dozen cases were all infected from the same source, especially if a known carrier is suspected to be responsible. Any major public health laboratory now has available a set of bacteriophages each of which will lyse only particular strains of typhoid bacilli. Typhoid bacilli can in fact be sorted into about twenty groups on the basis of which phages can lyse them. If we find for instance that all the six cases are of the same type, say A, we can be prepared to find they had a common source of infection, but if the suspected carrier is excreting type B then he has clearly no responsibility for the outbreak.

Similar epidemiological use of phage susceptibility is also standard practice in regard to such occurrences as outbreaks of staphylococcal skin infections in a nursery for premature infants. A harmless nasal infection in a nurse is often responsible and since virtually everyone has staphylococci in the nose it is important to pick the unwitting culprit correctly.

We can neglect all the magnificent research that has been done on the biochemistry and genetics of the bacterial viruses as irrelevant to our main theme but a little must be said about the variety of bacterial viruses that exist in nature. They are almost as multiform as animal viruses. There are the relatively large and complex DNA viruses of *E. coli* with which most of the classical research was done. They are tadpole-shaped with a hexagonal head and an almost unbelievably intricate mechanism in the tail for attachment to the bacterial wall and injection of DNA (Fig. 7). At the other end of the scale are tiny RNA phages so simply composed that we are promised the complete chemical specification of one of them in a year or two. Size and shape vary widely between these extremes and we should remember that for practical purposes there are very few really well known phages other than those which are parasitic on *E. coli*. Little is known – because virologists and bacteriologists have

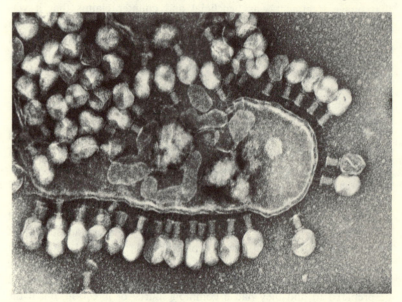

Fig. 7. Death of a bacillus: bacterial virus particles attached to the bacterial cell wall by fine tail filaments. (Electron micrograph from Dr L. D. Simon & Dr T. F. Anderson, *Virology*, **32**, 1967. Copyright Academic Press.)

not had the incentive to join forces and look – about the part played by bacteriophages in nature. A hint of what might be found is given in electron micrographs of the contents of a sheep's rumen. We see a multitude of bacteria, spirochaetes and protozoa with bacterial viruses of many shapes, free or attached to and attacking the microorganisms, as well as fragments of bacteria already disrupted by phage. There are apparently even more viruses than microorganisms in the rumen and what is happening there on a micro-micro scale is like a nightmare vision of an overcrowded world.

There are viruses structurally resembling bacteriophages which parasitize fungi and blue-green algae, and there are even tiny 'defective' viruses which can only live by parasitizing cells in which a 'helper virus' is already multiplying. In a sense this is a case of a virus parasitizing a virus, a situation clearly foreshadowed in Swift's famous verse:

> So, naturalists observe, a flea
> Hath smaller fleas that on him prey;
> And these have smaller fleas to bite 'em
> And so proceed *ad infinitum*.

All of which leads to that still unanswered biological question of how viruses evolved.

The first naïve assumption might well be that as the simplest of living things viruses must also be the most primitive, i.e. that they are likely to be derived from similar 'organisms' dating back to the dawn of life. This is wholly inadmissible. From its very nature a virus must be a parasite in the most fundamental sense, making use of the host's synthetic abilities for all its needs. There are only two plausible possibilities, or perhaps some combination or sequence of the two. The one most in line with ordinary biological experience is that viruses arose by a process of parasitic degeneration. Probably no microbiologist doubts that, as we discussed in Chapter 3, bacteria which became highly adapted to a parasitic existence gave rise to those somewhat virus-like organisms, the rickettsiae and the chlamydiae.

The rickettsiae are nearly all transmitted by biting insects, mites or ticks, and seem to have evolved primarily in these invertebrates. In many insects there are structures in certain cells which appear to be undoubted bacteria. They have not been grown artificially, and they are passed on to succeeding generations through the egg. What part they play, if any, in the animal's economy is unknown. Some of the smaller types closely resemble rickettsiae, and it is possible that the pathogenic forms have developed from these harmless parasites (or symbionts) of insects and ticks. The rickettsia of Rocky Mountain spotted fever shows a specially close resemblance to these forms. It is harboured by a tick which is quite unaffected by its presence. So mutually adapted are parasite and host that the rickettsia may be transmitted through the egg from generation to generation. The typhus fever rickettsia is transmitted by the louse, but in this case the adaptation of the insect host and parasite is much poorer. Infection of the louse with typhus is invariably fatal within a fortnight, evidence, according to Zinsser, that the louse has only recently taken on this role of vector.

65

The chlamydiae represent the next step and the organism responsible for the parrot (and human) disease psittacosis can be taken as the most convenient example. We have already used that condition to introduce in Chapter 1 the idea of a well balanced host–parasite relationship. The infective particles, even in their smallest form, are easily seen with the microscope, and under certain circumstances much larger forms appear. Like the rickettsiae they have both DNA and RNA and are susceptible to several of the antibacterial drugs. For these reasons they are not now regarded as viruses but placed with the rickettsiae as degenerate bacteria.

To derive the viruses from such forms we have to press the idea of parasitic degeneration to its absolute limit. There are analogies at higher levels of organization where it is characteristic of many though not of all parasites that the organism will tend to evolve in the direction of losing all unnecessary faculties, not infrequently being reduced until it is little more than a simple mechanism for providing sufficient nourishment to the reproductive cells. There are several well-known examples that adorn all the text books. The most spectacular is perhaps the Malayan plant *Rafflesia,* which produces the largest flower in the world – five feet in diameter. As a flowering plant, *Rafflesia* must have evolved from some fully functioning green plant, but it is now so intensely specialized as a parasite that it has no leaves, stem or roots. It is a mere network of feeding tubes which infiltrate the substances of living trees, abstracting sufficient nourishment to produce its extraordinary efflorescence. In the animal kingdom, we find the rather similar example of *Sacculina,* a parasitic crustacean which in its adult stage is reduced to a sac-like reproductive organ nourished by slender tentacles extending into the host's tissues. Less extreme examples are to be found in every group of parasites, and on a smaller scale it is what might be assumed to have happened in the evolution of viruses from pathogenic bacteria.

Bacteria have their own complex biochemical mechanisms, genetically controlled, by which their needs for substance and energy are fulfilled. Inside a living animal cell there is a basically similar set of mechanisms for the cell's own nutrition. If a parasitic bacterium could utilize the cell's metabolic equipment for its own use it could in principle survive the loss of many of its own genes. A corresponding reduction in size would become possible, and it is conceivable that such a process might continue until the organism is reduced to the absolute minimum, a fragment of matter still retaining the power of growth and multiplication, and capable of surviving in its own particular biological environment, but otherwise 'sans teeth, sans eyes, sans taste, sans *everything*' – just a single molecule of nucleic acid carrying only enough genes to provide for its own replication and for a protective coat of protein needed to convey it safely from one host cell to another.

The second hypothesis is that viruses represent the descendants of cell components that have gone wrong, mutinied as it were, to attack their host cell and then, as is inevitable with any self-replicating unit, become independently susceptible to the processes of evolution. In the early days when we knew virtually nothing of viruses this hypothesis was rather popular and the description of viruses as 'footloose genes' was often used. Then it became so unfashionable that in the 1940s it was legitimate to entitle a book *Virus as Organism* and come down firmly on the side of parasitic degeneration. Now that we know a great deal particularly about bacterial viruses the wheel has nearly come full circle. There are phages which can incorporate their DNA into that of the host bacterium so intimately that both sets of genetic material duplicate in unison. To all intents the phage DNA has become a part like any other part of the bacterial DNA – bacterial genes and viral genes have become indistinguishable. This control and cooperation can however be broken by appropriate types of stimulation. The phage DNA reestablishes its autonomy, actively destructive virus particles develop and the host cell is disintegrated.

If one attempts to look broadly at the immense range of genetic phenomena that have been detected in bacteria it is quite easy to accept the suggestion that the bacterial DNA is subject to an almost infinite range of possible accidents and in view of the almost infinite population numbers that bacteria can attain, could well give rise to a DNA fragment with sufficient autonomy to make it a possible ancestor to a line of bacterial viruses. There is experimental evidence of significant resemblances between the DNA of some bacterial genes and that of bacteriophages, to support these ideas of evolutionary relationships. It becomes intriguing to speculate on how many bacterial genes are in fact integrated viral genes and whether such genes were themselves of bacterial origin in the first place!

The suspicion that similar things could happen amongst mammalian cells is also growing. At a different level, a cancer cell is derived from an initial cell which instead of conforming to the normal intercellular discipline of the body, threw off all constraints and became essentially a parasite. Perhaps similar things can happen at the subcellular level. A curious hint has come as it were in the reverse direction by the growing acceptance in the last few years of the view that the mitochondria of all typical eukaryotic cells and the plastids of green plant cells both represent the incorporation into higher cells of primitive prokaryotic cells equivalent to bacteria and blue-green algae respectively. Both are now integrally controlled parts of the host cells but both have kept their own DNA and some vestiges of autonomy. No clear example of the liberation of a virus from an animal cell certifiable as 'uninfected' has ever been established. It is not the way of things that significant steps in Nature – things which, needing only to happen once, could choose any one of a thousand million occasions – should occur to order in a laboratory. Long ago herpesvirus was thought to have such an origin by no less a man than Doerr and today the nature and origin of the disease scrapie is so obscure that it has given rise to rather similar speculations. They envisage the possibility that, owing to some genetic anomaly in certain breeds of sheep, there is always a chance that one of its brain cells will develop a molecule or some simple collection of molecules with what can be called an autocatalytic power. If such a unit moves into an adjacent cell it stimulates the cell to produce a new crop of the same autocatalytic units. Once such a process were set in action, and provided damage were done to the function of the stimulated cells, the condition could not be distinguished from a true virus disease. Such a speculation is little more than a rationalization of three facts about scrapie: that it is limited to certain breeds of sheep, that it is transmissible by inoculation and that no recognizable virus particles containing nucleic acid have been demonstrated. We shall only be in a position to pass judgement on such an hypothesis when the responsible agent is chemically definable.

Many human diseases are caused by viruses but it is quite impossible to say *how* many. It should be clear from what has been said already about subclinical infections and the capacity of viruses to vary in virulence from strain to strain that no figure can be given for the number of viruses that are pathogenic for man. Every year or two we hear of an epidemic produced by a virus formerly unknown or known only as an unimportant agent occasionally isolated perhaps when mosquitoes were being surveyed for virus carriage or intestinal viruses being sought in infants. A case in point is the Lassa virus, first isolated in 1969, which produced fatal infections in a mission station in West Africa and in laboratory investigators studying the new virus in the US.

Some attempt must however be made to give an overall impression of the importance and general character of the main virus diseases of man. All things considered, what seems expedient here is to list those viruses which within the last seventy years have been responsible for widespread easily recognizable disease in substantial numbers of people. We shall list them under the generic names given in the table at the beginning of this chapter with no more than a note or two on points of current interest in 1971.

Poxviruses. Smallpox and its lesser variants are now largely confined to tropical areas. Its extent is diminishing fairly steadily and if the developing countries are allowed to develop sound health services smallpox could be the first of the plague diseases to be eradicated.

Herpesviruses. Herpes simplex, characteristically of the lips, is very commonly seen in all parts of the world but is becoming less frequent. Chickenpox and herpes zoster (or shingles) are caused by another virus of this group. With only minor reservations the growing opinion that infectious mononucleosis (glandular fever) is in part or wholly due to a herpesvirus can be accepted. The EB virus isolated from certain tropical cancers may not be the cause of those diseases but has turned out to be an extremely common and usually harmless inhabitant of millions of people everywhere.

Myxo- and paramyxoviruses. Here we find influenza, measles and mumps. A relatively recent newcomer not yet known to the layman is respiratory syncytial virus which is probably responsible for more cases of severe bronchitis and pneumonia in infants than any other agent, while the parainfluenza viruses are the most important cause of 'croup'.

Rhabdoviruses. Rabies (hydrophobia) may be acquired by bite of a 'mad dog', bat or occasionally other animals. To our knowledge there is only one reported recovery from established human rabies. In 1967 a previously unknown rhabdovirus, the Marburg virus, killed seven laboratory workers handling African monkeys which had been sent to Marburg in Germany.

Arboviruses. These *ar*thropod-*bor*ne viruses, carried by mosquitoes, ticks or other arthropods, represent a large and heterogeneous group of over 200 viruses which may be split into natural groups with appropriate names. Yellow fever is the prototype, while dengue fever is perhaps the most widespread. Local varieties of rather closely related viruses are responsible for severe or fatal encephalitis or haemorrhagic fevers in various tropical and subtropical regions.

Picornaviruses. The three types of poliovirus are the humanly important examples but this very large group also includes thirty-three echoviruses and thirty coxsackieviruses which occasionally cause illness following infection of the intestinal or respiratory tracts. Another subgroup comprises the hundred or so rhinoviruses responsible for most cases of the common cold.

6. Infection and immunity

In a real sense all the medical sciences that impinge on infectious disease – bacteriology, virology and immunology – take their origin in the folk knowledge that a man with a face pocked by smallpox would never again be subject to that disease. Immunity followed an infectious disease and was specific. Jenner added the first practical application and a new slant to the concept when he showed that a similar but manifestly different 'principle of disease', which we now know as Jennerian cowpox virus, could harmlessly immunize against the more serious disease. When bacteriology became a science in Pasteur's hands his first attempts to apply it to the prevention of animal and human disease were conscious efforts to 'attenuate' the infection so that it was harmless yet could immunize in the same fashion as Jenner's cowpox.

This set the pattern and at least until 1920 the study of immunity was almost exclusively related to infectious disease. The applications of immunological techniques and ideas progressively widened. Between 1886 and 1890 the use of killed bacterial vaccines for immunization, the recognition of the capacity of blood to kill anthrax bacilli and the production of antitoxin against the toxins of diphtheria and tetanus, gave a tremendous stimulus to immunological research. All this activity was nominally at least related to infectious disease but in 1898 there came a major turning point in the history of immunology when Bordet found that rabbits could be immunized against red blood cells from another species, producing antibody which could specifically dissolve (haemolyse) the red cells. Immunology became immediately a biological science in its own right instead of a mere approach to the understanding of infections.

Particularly since 1940, academic immunology has moved far away from its original field and in the process a wholly new experimental and theoretical approach has been developed. Today we are almost compelled when writing about the application of immunology to infectious disease to reverse the historical method. Instead of describing how findings in infectious diseases allowed the development of a science of immunity, our preference is to outline the current concepts of immunology and show how they throw light on the phenomena of infection and facilitate its control.

Completely central to modern immunology is the concept of antibody and immunoglobulin. The terms are synonymous, 'immunoglobulin' being used when we are concerned with it as a chemical molecule, 'antibody' if the specificity of its reaction with antigen is what interests us. At the chemical level an immunoglobulin is a relatively complex protein of which there are at least five types, G, M, A, D and E, but we need at present only consider the commonest and simplest type, G, now regularly symbolized as IgG. This is a protein of molecular weight around 155,000 and made up of four chains of amino acids connected by disulphide bonds as shown in Fig. 8. It is a symmetrical molecule with two functionally active ends. At each end the terminal segments of both 'light' and 'heavy' chains cooperate to form a potential combining site. The two ends are exactly equivalent chemically and functionally and give the immunoglobulin molecule its function as an antibody.

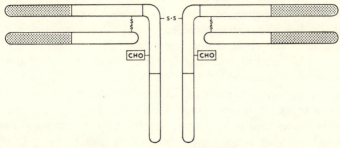

Fig. 8. Modified Porter diagram of an immunoglobulin G molecule. To show symmetrical structure with two 'heavy' chains, each of four subunits, and two 'light' chains (with two subunits) joined by S–S (disulphide) links. Each light and heavy chain has a 'variable' subunit (stippled). The juxtaposition of a light and a heavy variable segment at each end of the molecule provides the combining sites for antigen.

It is unnecessary to discuss the problem of how the body manages to provide a quite extraordinarily wide diversity of antibodies. All that need be said is that the pattern of an antibody, like that of any functional protein, is determined by genetic processes and is expressed in the stem cells from which all cells concerned in the business of immunity are derived. Once established, the pattern remains constant and is transmitted to all descendants of the stem cell. 'Clone' is the biological term for a population derived by simple non-sexual proliferation from a single organism or cell; all members of one stem cell clone are immunologically uniform. The physical basis of the specificity of an antibody resides in the fact that the terminal segments which make up the combining site have unique sequences of amino acids distinct from those of other

71

antibodies. It is of the essence of modern immunological theory that the diversity of antibody pattern is determined genetically, *not* as a result of contact with different antigens.

Antigen must now be defined in terms of antibody. An antigen is a substance which by virtue of a particular chemical grouping (an 'antigenic determinant') can unite specifically with the combining site of an antibody molecule. There is another common usage of the word 'antigen' as something which will stimulate the body to produce a corresponding antibody, but 'immunogen' is the better name for such a substance. When a particular immunogen makes specific contact with the corresponding antibody carried as receptors on the surface of a small proportion of the body's lymphocytes it stimulates these particular cells to proliferate and, when conditions are appropriate, to synthesize the corresponding antibody. Most antigens can also function as immunogens and in general most proteins and complex glyco- or lipoproteins are both antigens and immunogens. There is, however, one very important limitation. For all practical purposes no normal component of the body can act as an antigen or immunogen in that particular individual. This, which can be taken as axiomatic, is referred to as natural, or intrinsic, immunological tolerance.

It is convenient to speak of all cells which carry antibody-type receptors as immunocytes, although the term antigen-reactive cells is more commonly used in the laboratories. The standard structural form of most immunocytes is the lymphocyte and at least 50 per cent of all lymphocytes in blood, spleen or lymph nodes are immunocytes.

It has become clear within the last five years that immunocytes fall into one or other of two categories. They are spoken of as thymus-dependent cells (of the T system) or thymus-independent cells (of the B system) for reasons which need not be elaborated. It is the functional difference between the systems which alone concerns us.

When an immunocyte of the T system is stimulated by the corresponding antigenic determinant of an immunogen it proliferates to produce more cells like itself. All descendants carry the same antibody receptors but they do not liberate free antibody. Such T cells are directly involved, as cells, in those immune reactions for which they are responsible. They provide the means by which foreign tissue grafts are rejected by the body and mediate such skin reactions as the Mantoux test for sensitivity to tuberculin in patients who have suffered from tuberculosis. As we shall see, the T system has significant activity in relation to viral and fungal infections.

The B system, also called the plasma cell system, is responsible for all or very nearly all liberation of antibody into blood and other body fluids. A lymphocyte of the B system, when specifically stimulated, begins to proliferate and in the process changes its structure toward that of a plasma cell. In essence a plasma cell is a lymphocyte which has converted itself into a highly active factory for synthesizing its own genetically determined type of antibody and setting it free into any surrounding fluid. Antibody has a number of functions in dealing with infection, particularly bacterial ones. Within the B system there are cells which produce one or other of three common and two rare sorts of

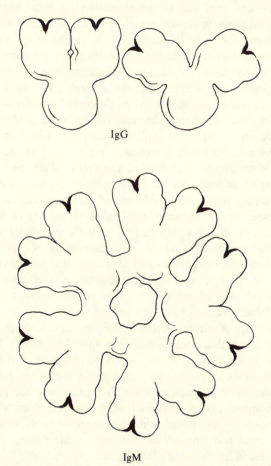

IgG

IgM

Fig. 9. Semi-diagrammatic sketches to show the probable shape of molecules of IgG and IgM. Note the flexibility of the molecule shown in two configurations for IgG. The IgM molecule is effectively a pentamer of IgG but there are some additional differences.

73

immunoglobulin. IgA has the special virtue of being liberated on internal surfaces like the lining of the bowel or of the respiratory tract and is instrumental in protecting these mucous membranes, for a time at least, against reinfection. In the blood, IgG is much the commonest type but IgM is also important; the molecular weight of IgM is more than five times that of IgG and functional differences between the two reflect this size difference (Fig. 9).

*

In rather concentrated and simplified outline this is the immunological equipment of a mammal like ourselves. We must now ask how it is relevant to our theme of the behaviour of infectious disease either in some natural ecosystem or in the sophisticated environment of preventive medicine and hospital care. The obvious first step is to go right back to the beginnings of medical speculation and ask what modern immunology has to say about how patients recover from infectious disease and why they are subsequently immune to that infection. Three diseases will provide representative examples: pneumonia, which is a classical bacterial disease, diphtheria, where toxin produced by the bacillus is the disease-producing agent, and measles, a typical viral disease.

Pneumonia was once very much more important as a cause of death than it is today and in the 1930s one of the great central themes of medical research was the understanding of all aspects of lobar pneumonia. Much of the work was magnificently done and still provides us with the clearest picture of how a bacterial infection is overcome. With the advent of sulphonamides and penicillin the natural process of recovery was pushed into the background and in looking at the immunological side of recovery we are automatically committed to looking at lobar pneumonia as it was seen in the 1930s.

In the absence of effective antibacterial treatment lobar pneumonia is a severe disease with an overall mortality up to 33 per cent. The patient is extremely ill and remains so for six or seven days. Then, if he is one of the fortunate ones, there is a dramatic turn for the better. His temperature falls rapidly to normal, delirium disappears and the patient feels weak but infinitely better. Obviously something important was happening at the time of the 'crisis' and its interpretation in immunological terms has remained one of the foremost achievements of medical research in the fullest meaning of that term.

Basically the situation was simple. There are many different types of pneumococci, something like eighty all told, but only a dozen that are or were common causes of pneumonia. They are differentiated immunologically in terms of their major antigen, a gummy polysaccharide that forms a capsule which completely envelops the organism. This became known quite early as 'specific soluble substance' (SSS) and it is convenient to use that abbreviation still. There are almost as many patterns of complex polysaccharides as there are of proteins and each type of pneumococcus has a pattern distinguishable from that of any other type.

The SSS is a good immunogen and a wide variety of immunological tests can be used to show whether there is anti-SSS in serum. The most illuminating is the mouse protection test. It is easy enough to find a strain of pneumococcus type I which is highly virulent for mice. Inject a standard dose of say 10,000 pneumococci into the body cavity and the mouse dies in two to three days. Mixed with a serum containing antibody to SSS–I and similarly injected, that same standard dose is quite harmless. But go one step further and add to that harmless mixture some pure SSS–I and it becomes fully lethal. Antibody against SSS is what protects the mouse – it works by combining with SSS in the bacterial wall and so coating each pneumococcus with a layer of globulin. Such coated ('opsonized') organisms are rapidly taken up ('phagocytosed') and destroyed by the scavenging cells of blood and tissues.

The situation in the mouse can be applied almost directly to the patient with pneumonia. When he is in the first stage of his illness the pneumococci are multiplying rapidly in the lung, and their enzymes are damaging both the respiratory membrane of the lung and its blood capillaries. Fluid and blood cells enter the airspaces and produce the dark consolidation of lobar pneumonia. There are many phagocytic cells – the polymorphonuclear leukocytes of the blood and the lung's own local scavengers – but the amount of SSS being produced by the pneumococci mops up any antibody in the region before it can coat more than a small fraction of the multiplying pneumococci. The fate of the patient is determined by the speed with which he can make enough antibody to saturate the protective screen of SSS and still keep sufficient to coat the actual pneumococci. If the pneumococci can continue to produce more SSS than the body's antibody can deal with, the organisms will go on multiplying and produce more SSS in a vicious circle that will end with the patient's death. But once antibody gains control all the processes will now work in favour of the patient.

The pneumococcal SSS is a curious antigen in that it is almost incapable of stimulating the T system of immunocytes. The immune response is almost wholly due to antibody and again it is unusual in that IgM antibodies are predominantly concerned.

If the 'crisis' and natural recovery from pneumonia is due to the development of an adequate level of antibody, this should have two important implications at the practical level: first, that appropriate antiserum against the right type should be capable of curing pneumonia, and second, that vaccination, again with the right type of SSS, should protect against the disease. Within limits both turned out to be true but almost at the very moment when technical difficulties in applying them were being overcome, sulphonamides active against the pneumococcus were introduced. Sulphapyridine and sulphadiazine were so much simpler to handle than serum treatment, and at least as effective, that interest in the serum treatment of pneumonia vanished almost overnight.

Immunization against pneumonia has been tried and found to work in two populations known to have an exceptionally high incidence, Bantu labourers in South African gold mines and army recruits in the United States. The last and most convincing test was done in 1945 in an army camp in South Dakota. In previous winters there had been relatively large numbers of cases due to six types, I, II, IV, V, VII, and XIV, and it was decided to produce and test a vaccine containing types I, II, V, and VII, omitting the other two common types, IV and XIV, in the hope of obtaining evidence as to whether protection was specific, i.e. active only against types of pneumococci that were included in the vaccine. During the winter season all cases of pneumonia were checked as to the type responsible and the figures compared with those of previous years. At the end of the winter it could be said that pneumonia due to those types that were included in the vaccine was only 17 per cent of what it had averaged in the previous winters, while there had been no reduction in pneumonias due to types IV and XIV. The results were excellent, but with penicillin available there was no social or military reason to continue.

Before we leave pneumococcal infections as a type in which almost the whole pattern of disease is determined by one bacterial antigen and the corresponding antibody, a new topic must be introduced, immune deficiency disease. Paediatricians had always known that some infants failed to thrive, had repeated attacks of pneumonia and died, usually when they were about a year old. With the appearance of penicillin more of these children survived and in 1954 it was discovered that some such pneumonia-prone infants had no immunoglobulin in their serum and could make no antibodies after being given the usual vaccines. In those days immunoglobulins were called gamma globulins and the disease has continued to be known as agammaglobulinaemia. It is a sex-linked genetic disease due to an altered gene on the X chromosome and seen only in male infants. Many cases have since been found and studied. Obviously their response to infectious disease has been a matter of special interest. As far as pneumococcal pneumonia is concerned the story is just what would be expected. At birth the baby is protected by anti-bacterial immunoglobulin G transmitted across the placenta from its mother. As that passively acquired antibody is gradually destroyed over the ensuing six months or so, infections become progressively more frequent and severe but they can be controlled by penicillin and prevented by a regular injection of pooled human immunoglobulins from normal donors. Sometimes the children do well. A year or two ago it was reported that the very first baby in which agammaglobulinaemia was diagnosed is keeping well on a monthly dose of globulin – and has an excellent job as a bank teller in West Virginia!

Lobar pneumonia cannot be taken as a model for all bacterial diseases. Every one of them has its own special qualities and the host's response is correspondingly varied. Almost every other group of pathogenic bacteria will provoke responses from the T system of immunocytes as well as the production of antibody. In tuberculosis, sensitization to tuberculin dominates the picture both in regard to the progress of the disease and for diagnostic tests. Sensitization can be interpreted as due to a vigorous production of T immunocytes which on contact with the corresponding antigen suffer damage and release chemical substances with drug-like actions on other cells. In the individual infected with the tubercle bacillus such contact with tuberculin produced within the lesions in the lung may play an important role in the development of the disease. Similar reactions follow the introduction of small amounts of tuberculin into the skin in the form of a Mantoux test. The resulting changes in capillary blood flow produce the dull red reaction of a positive skin test that indicates past or present infection with the tubercle bacillus.

In another group of bacterial infections the picture is dominated by the production of a dangerous toxin by the organism. Diphtheria is the classical example (see Chapter 14) but tetanus is equally important. Many, perhaps most pathogenic bacteria produce soluble proteins which are toxic to laboratory animals and may sometimes manifest their action in man. Scarlet fever, for rather obscure reasons, has almost vanished but like diphtheria it was a throat infection whose general symptoms were due to a toxin produced by the responsible streptococcus.

At the immunological level the important feature of the diseases for which toxin producers are responsible is the key role played by antibody (antitoxin) which can inactivate the toxin. The use of antitoxin made in horses to cure the disease and of immunization with modified toxin as a preventive measure are discussed in relation to diphtheria in Chapter 14. Similar principles are applicable to tetanus.

When we come to the viruses we find a rather different pattern of immune response during the process by which recovery takes place and long-lasting immunity is established. Before discussing immunological aspects however some mention should be made of intracellular defence against viral action by a naturally occurring substance known as 'interferon'. Interferon is a protein of low molecular weight which is synthesized and liberated by cells in the course of infection by viruses of many types. The protein readily diffuses through the fluids of the body and into uninfected cells which it can render insusceptible to virus damage. There is much current interest in the fact that viral nucleic acid, or even artificially synthesized double stranded RNA, will stimulate an animal to make large amounts of interferon and protect it against infection by any of a wide variety of viruses. The possibilities of such RNAs for treatment are obvious but there are still certain practical difficulties in the way. We are still ignorant of much of the process by which interferon is produced in the cell and of its evolutionary significance. There is a good deal to suggest that the virulence of a virus is inversely related to its capacity to provoke interferon production.

Whatever may be the eventual interpretation of the role of interferon in recovery from viral infection there can be no doubt that immune responses play an important and sometimes rather surprising role in the development of some virus lesions and in the process of recovery and persisting immunity. Again we must add the warning that immune responses and their effects will differ according to the virus being studied. Measles is the most familiar of all virus diseases and perhaps provides the most illuminating example of the role of immunity in virus disease.

Once a child has had a typical attack of measles he will never have another attack. This is a fairly general rule for any of the acute viral infections in which the virus spreads around the body via the blood. For many years it was taught that post-measles immunity was due to the continuing presence of antibody in the blood. On this assumption the use of normal human globulin was successfully introduced as a means of protecting sickly or otherwise vulnerable children who had been exposed to measles. When it became possible to measure antibody against measles in the laboratory it was found as expected that antibody persisted in the blood for many years after the immunizing attack.

With the recognition of considerable numbers of children with agammaglobulinaemia and their maintenance in fair health by immuno-globulin injections, it was inevitable that some of them would contract measles. To everyone's surprise they showed a normal measles course with a typical rash which faded at the normal time and was followed by just as substantial immunity against reinfection as would be shown by any other convalescent. Antibody production is therefore not *necessary* either for recovery from or for the development of immunity to measles. If, as everything suggests, the child with agammaglobulinaemia has no functioning B immunocytes, the immunity he acquires after measles must depend solely on the T system of cell-mediated immunity. This makes it of special interest to look at what happens with measles in a child whose T system is not functioning. There is a condition of congenital absence of the thymus gland which results in complete absence of T immunocytes, but these infants are very rare and very frail. None has lived long enough to catch measles. However when children with acute leukaemia are under standard treatment with cortisone the combination of disease and treatment virtually paralyses the T system of immunity. Such children are in great danger if measles or chickenpox appears in their ward. Measles usually kills these children, but in quite abnormal fashion. There is no rash and the children die of what is called 'giant-cell pneumonia' in which the lung is choked with large cells containing many nuclei. One or two of these cases have survived for some weeks and for the whole period measles virus is present in their lungs. The normal process of recovery and elimination of the virus is not occurring.

Taking both of these abnormal conditions into consideration we are almost driven to believe first, that the rash of measles is a T system immunological response, and secondly, that for there to be persisting immunity the T system must function. The first conclusion was actually suggested by von Pirquet in slightly different form many years ago from a consideration of the nature of the incubation period in measles. As everyone knows it is usually nine or ten days after a child has been exposed to measles before he shows the early catarrhal symptoms and another two or three days before the skin rash is well developed. It is worth thinking about what is happening during those two weeks.

The first point to be made is that measles is not intrinsically a cell-damaging virus. When it is grown in cell culture the most conspicuous result is that groups of cells fuse together. The intervening cell surface membrane disappears to produce giant cells with many nuclei. It seems that during the first ten days in the body measles virus behaves similarly. It gets in presumably through a little focus of infection in the respiratory tract, moves to the local lymph nodes and multiplies in lymphocytes and is thereby carried to all parts of the body. No symptoms are produced at that stage but it is known that in the lymph nodes and spleen giant cells like those in our cell cultures are becoming common. Soon the virus and cells carrying it are abundant in all the collections of lymphocytes in the body, i.e. in spleen, lymph nodes and the lymphoid tissue along the gastrointestinal tract from tonsil to appendix. These however are also the sites where immune responses are generated. Virus goes on multiplying but very soon cells tuned to react to measles antigens are also proliferating as T immunocytes and several varieties of B cells. When a T immunocyte meets its corresponding antigen in more than minute amount its surface is damaged and it liberates a variety of drug-like substances which in their turn can be highly damaging to adjacent cells. For instance, if a fully developed T immunocyte tuned to react with measles antigen makes contact with a cell in which measles virus has been quietly growing, the encounter will be disastrous for both. The important result will be that the cell containing the measles virus is disintegrated. During the incubation period therefore we have building up in the body two populations of cells violently inimical to each other. As long as they keep apart nothing happens, but with rising populations this becomes impossible and once confrontations begin to occur much more viral antigen is set free which in its turn can 'fire off' more T cells. So the situation builds up into a runaway explosion and wherever T cells and measles cells lodge and react there will be damage to capillaries and leakage of fluid into the tissues. When that occurs in the skin we call it a measles rash.

The B cells are active too and almost certainly are influenced by the reacting T cells. At the height of the rash the blood contains an unusually large number of plasma cells and antibody is being rapidly formed. Aggregates of antigen and antibody can also be irritating to tissues and may play some part in producing the spots.

What still remains obscure is just how the mass of virus is inactivated and eliminated both in the normal individual and, much more so, in the individual with agammaglobulinaemia. All that we can suggest is that in the presence of large numbers of T immunocytes any cell in which measles virus is beginning to multiply will almost certainly be destroyed by contact with T-immune cells before a brood of descendent virus particles can develop.

Measles may be a familiar disease but it has two important aspects which have only recently been recognized. The first is that about the time of the rash, inflammation of the brain, encephalitis, of widely variable severity may occur. Some cases are fatal, others show no more than an abnormal tracing on the electroencephalogram. Such minor brain involvement is quite common. Probably all the cerebral signs and symptoms are also of immunological origin; they may well represent the equivalent in the brain of the rash on the skin. What is worrying to paediatricians is the suggestion that these episodes can sometimes do subtle harm to the personality or educability of the child. That risk has been one of the important reasons for supporting universal immunization against measles.

Finally, only four years ago a rare brain disease of children and adolescents with the forbidding name of subacute sclerosing pan-encephalitis (SSPE) that had been known for many years, was found to be a late sequel of measles. It is an extraordinary condition in which measles virus keeps on slowly spreading from cell to cell through the brain. There is an active and continuing B type immune response. The brain itself is peppered with plasma cells and there is antibody everywhere, including the cerebrospinal fluid. The B response is obviously of no avail because this insidious disease is invariably fatal, sometimes years after the original measles attack. We can only speculate that somehow or other these children have lost those particular T cells which can react with measles antigen, or that the anatomy and physiology of the brain is such that any anti-measles T cells in the body are impotent to influence it. SSPE is very rare and we do not know how it arises but it has left a nagging question in the minds of the makers of live attenuated measles vaccines. Will the vaccine strains ever produce SSPE? The answer will probably be no, but it may take some years before we are sure.

This account of immune reactions in measles would probably hold with only minor modifications for all the 'generalized' virus diseases. There are many other patterns of infection and the part played by immunity differs considerably from one to another. Few have been as closely investigated as measles and for most of them there is much information about antibody production and function with almost nothing known about the functioning of the T system. There are hints in several however that T cells may have something of the same two-faced 'Jekyll and Hyde' importance that they have in measles.

*

Still holding on to our objective of providing only that outline of immunology that is relevant to our central theme of the ecology and

81

control of infectious disease, we come to the all-important topic of immunity without disease. Perhaps it betokens a soft-hearted, simple-minded approach to say that, on the whole, Nature prefers that neither host nor parasite should be too hard on the other. For Nature, survival of the species is all that counts and the norm, if there is one, is that the host should not die and that his infection should be passed on to one new host individual. Here subclinical infection and natural immunization are all-important.

Before moving on to the main topic, however, we should mention a new facet of immunology that is appearing in contemporary technical discussions. With the acceptance of genetic, including somatic genetic control of antibody pattern, people are beginning to ask just what is the basic range of antibody patterns whose specifications are fused from the information in sperm and ovum at fertilization. The obvious answer that they are those patterns whose ready availability best makes for species (and on the average, individual) survival still begs the question. At present there is no obvious way of finding what are the most important immunological needs for survival. We might guess that the answer was to ensure the harmlessness of the vast numbers of bacteria in the intestine. There is a hint or two in that direction but they could be quite misleading. It represents a fascinating problem for the next decade and that really is all that can be said.

Whatever the *fons et origo* of the diversity of immune capacity that the infant possesses soon after birth, we know that it is adequate to respond to most of the potentially pathogenic microorganisms of his environment. Twenty or twenty-five years ago there were field epidemiologists busy taking blood samples from slum children in cities like Cairo, Calcutta or Manila, big enough samples of children to allow a statement of what percentage had this or that antibody at each year of age. Most of the investigators were concerned with polio but when the samples reached the base laboratory in America or elsewhere it was easy to test them for antibody against other sorts of virus. The pattern of the results was simple. By the time children were three years old most of them were immune to every common pathogen in their environment. The converse of this can be seen when large numbers of young adults from country districts in relatively empty and prosperous lands – the United States, Canada, Australia – are brought into crowded recruit camps at the outbreak of a major war. They have had little experience of infectious disease and their sera contain few antibodies. They spend their first winter in camp having a succession of infections and they are often severely hit. Measles killed many young soldiers in the American Civil War.

The reaction of very young children to infectious disease is rather paradoxical. The figures for any community with well developed medical statistics will always show that the greatest incidence of deaths from well defined and common infectious disease is in the very young. This holds for measles, whooping cough and all the common types of meningitis. On the other hand there are several diseases where the important symptoms of a disease when it occurs are much more severe in adolescents and young adults than in infants and small children. Infectious hepatitis hardly ever produces jaundice in children under two; in cities where all children are naturally affected by polio viruses before they are three the incidence of paralytic polio in infants is nevertheless very low. There is a great deal to suggest, too, that yellow fever in its heyday was trivial in young European children but often lethal in young 'unacclimatized' adults. This phenomenon will be discussed more extensively when we deal with epidemiological aspects. Here we are only concerned to point out that in young children there can be an excellent antibody response to a wide variety of pathogens with relatively trivial symptoms. Equally there are several types of infection such as measles and polio which tend to be disproportionately severe when they affect a completely non-immune young adult.

It is interesting to look at these phenomena in the light of the two immune systems, T and B. If we omit the pneumococcal infections it is the general rule that any type of infection stimulates a response in both systems. Antibody is produced and so also are specifically tuned reactive lymphocytes of the T series. Judging from measles the T system is the more important in establishing long-lasting immunity but it is also closely concerned in the production of symptoms and the massive liberation of virus. In all probability the greater intensity of symptoms of many infections in non-immune young adults depends on the fact that the reactivity of specifically stimulated T cells, like other functions of the body, reaches a maximal level at around the age of twenty. Another factor which may turn out to be important is the interaction of the two systems. If, for instance, there is a high level of antibody against microorganism X in the blood when an infection by X occurs, antibody may coat the organism so rapidly that very few of its antigenically active sites have any chance of contacting a reactive T cell. Other more subtle aspects of the interaction of T and B systems are under study at the present time but they are not yet relevant to the process of infection and recovery.

We can summarize by saying that when a microorganism of infectious disease is common in a human community a fairly standard pattern of interaction between parasite and host usually develops. Children are infected early in life and a proportion, perhaps largely those with genetic inadequacies or suffering from malnutrition and lack of maternal care, will die. The great majority will be infected at an intensity which they can easily handle. Some infections, like measles and smallpox, will almost always show more or less typical symptoms. More frequently, the various infections that reach a child either produce no symptoms or give a clinically undiagnosable minor fever. Each infection, however, will find an opportunity to stimulate appropriate T and B immune responses. The level of immunity may sometimes be high enough to make the child completely invulnerable to reinfection. It is, however, probably more frequent to find that although immunity cannot wholly prevent reinfection, any subsequent episode is short-lasting and without symptoms, serving simply as a 'booster' to reinforce immunity.

Modern civilization with its concern for pure food and water, effective disposal of sewage and its support for education in child care and personal cleanliness, has been of the greatest help in diminishing infectious disease. But it has at the same time eliminated a large proportion of the subclinical infections that immunized young children harmlessly against infections which are more dangerous or more inconvenient if they are first met in adolescence or early adult life. Polio notoriously became a serious problem only as countries developed a high standard of living and civic cleanliness. Immunization by artificial means became a necessary aspect of life in the modern affluent society.

In 1971, immunization of children against diphtheria, tetanus, whooping cough and polio is almost universal. Measles immunization is now aimed to cover all children in the United States and its use is spreading in other countries. Effective vaccines against mumps and rubella are available and can be expected to come into general use within the next decade.

A second set of immunizations can be listed which in general will only be used under circumstances of special risk. In the countries of Europe, North America and Australasia, only travellers to other areas will in general be immunized against yellow fever, cholera or plague. Policies and recommendations vary in regard to vaccination against smallpox or immunization against typhoid and paratyphoid fevers. Sometimes laboratory workers must be protected against a dangerous organism which is being studied and something equivalent may be called for when limited regions suffer from the activity of an unusual pathogen.

Many of these vaccines are referred to in other chapters, and probably all that justifies discussion at the purely immunological level is the nature of the immunizing material to be used. With bacterial infections the standard approach has been to inject a suspension of bacteria killed by heat or a chemical agent such as formalin or alcohol. Typhoid and paratyphoid (TAB), cholera, pertussis (whooping cough) and plague vaccines are the examples in current use. In diphtheria and tetanus the bacterial toxin is solely responsible for symptoms or death and immunity against the toxin is all that is required. Toxoid, i.e. toxin rendered innocuous by formalin treatment, is always used, usually in the form of a precipitate with aluminium hydroxide. We have already mentioned the experimental immunizations against pneumonia in which the significant antigen of the pneumococci, the polysaccharide SSS, was used in pure form.

The controversy between those who favoured killed vaccines and those wishing to use living attenuated virus became public knowledge in the 1950s when Salk (killed) and Sabin (living attenuated) vaccines against polio competed for public favour. Both are still in use but it is probably significant that almost all recent developments of new vaccines against viruses make use of attenuated living virus. This, of course, is exactly what Jenner used in 1798!

Accepted vaccines in which living bacteria or viruses are used at the present time are:

1. Against tuberculosis: BCG (bacille Calmette-Guérin).

2. Against viral infections: Smallpox vaccine (vaccinia virus), Sabin polio vaccine, yellow fever vaccine and the current vaccines against measles, mumps and rubella.

The only real criterion for choice of living against killed vaccines is their relative effectiveness and safety in practical field trials. Broadly speaking the main advantage claimed for killed vaccines is safety. For several years there was resistance to Sabin vaccine because of fear that a living virus could mutate to a more virulent form. However, once its safety is established, a living vaccine presents several advantages. Probably the most important is that the immunogenic material, namely the protein coat of a virus, increases in amount in the body and is distributed to the same organs as would be involved in subclinical natural infection. One could expect, therefore, that an optimal amount of antigen will be produced and that the details of the immunizing process, e.g. the relative involvement of T and B systems, will be similar to that following natural infection. With a killed virus vaccine all the antigen must be supplied in the injection and in general it will pass only to a limited portion of potentially responsive sites in the body. There is no real information yet about the response of the T system to vaccination owing to lack of convenient quantitative approaches. It can hardly fail to be important and there are enough new experimental approaches being explored at the present time to suggest that some may soon be applied to the bread and butter problems of producing better vaccines.

The final and in many ways the most direct application of immunology to the understanding of the natural history of infectious disease is the rather prosaic one of providing diagnostic tests. Even with modern cell culture techniques the isolation and identification of a virus is a time-consuming business. Further, it can only be achieved if it is possible to obtain the relevant specimens within a short period of the onset of symptoms and if the material can be transported rapidly to the laboratory. If we are only needing to know whether a given fever is due to some virus we are currently interested in, by far the simpler approach is to take one sample of blood for serum as soon as the patient becomes sick and a second sample ten to thirty days later. If the first sample shows little or no antibody and the second a high level of antibody against the virus, then we have identified the disease. We need not go into great detail about the methods employed for identification of these antibodies.

Antibody against bacteria will usually be recognized by the power of serum to agglutinate or clump in easily visible aggregates a suspension of killed bacteria. An even more versatile test is complement fixation, but it is too complicated to explain in technical detail here. Rather simpler to follow is the principle of the neutralization test. One takes a standard dose of virus and adds a volume of serum. Then the ability of the mixture to produce visible infection in a cell culture, a chick embryo or some experimental animal, is tested. No infection means that antibody is present; if there is no antibody the virus will infect as readily as virus alone.

It is often of value to sample a cross-section of a community for the presence of antibody, as an indication of the proportion of persons at various ages who have already been infected by the virus or bacterium that we are interested in. The same methods are used but the results are sometimes difficult to interpret.

There are two ways of detecting evidence of past infection by using a skin test, i.e. the injection into the skin of a small amount of antigen. The Schick test for diphtheria immunity involves the injection of a small accurately measured amount of diphtheria toxin. If the individual has not experienced contact with diphtheria bacilli and therefore has no anti-toxin, there will be a dull red 'positive' area a couple of days later. If the subject has developed enough immunity and antitoxin to qualify as protected against diphtheria there is no reaction at all. It is essentially just a means of measuring whether a certain level of antitoxin is present in the blood.

The second type of skin test is used to show the presence of specifically reactive T immunocytes in the body and the prototype is the Mantoux reaction which when positive shows a slowly appearing reddened and swollen area around the point of injection. It is a typical delayed hypersensitivity reaction and signifies that the individual has at some period been infected, not necessarily seriously, by the tubercle bacillus.

7. Susceptibility and resistance

Every infection is in a sense a struggle for survival between host and parasite and when we look at the whole process from the point of view of the species concerned both must manage to survive. Only an infinitesimal proportion of the microorganisms resulting from multiplication in each host survive, but they are sufficient to ensure the continuance of the species. The host species on its side has developed during its evolution the normal defences against infection, and, as well, each individual during and after first infection by any particular microorganism develops some degree of specific immunity against it. This double mechanism of protection does its work well so long as the general efficiency of the body is maintained and so long as the attacking microorganism is not of abnormally high virulence, but sometimes it fails.

The reader will appreciate from all that has been said so far that 'infection' is not synonymous with 'disease' – indeed most infections are subclinical. So, in discussing susceptibility and its reciprocal, resistance, we must be careful to keep separate the factors influencing infection on the one hand and disease on the other. For instance, human beings are totally insusceptible to infection, clinical or subclinical, with myxomatosis of rabbits or distemper of dogs. Infection by polioviruses or arboviruses is common but only occasionally leads to clinically apparent manifestations of disease. In this chapter we shall discuss some of the factors – apart from specific acquired immunity – that determine whether an infected individual becomes ill or dies. For the most part we shall be concerned with the differences that are evident when different individuals are infected by the same strain of virus or bacterium. Why did only one child in a hundred of those infected during a prevalence of poliomyelitis become paralysed? Why can what seems a very ordinary staphylococcus sometimes find it possible to multiply almost without hindrance and kill the patient in a few days unless speedy and effective treatment is available? Ironically, despite immense investment in research we have only a rudimentary idea of the factors that determine the severity of infection in the patient with no prior experience of that particular microorganism. We can look at the factors under three general headings, inheritance, age, and the tissue or organ primarily involved. Immunity acquired after experience of a similar infection or artificial immunization is of course the most important factor of all in conferring resistance but here we are dealing only incidentally with immune responses. The differences that interest us in this chapter are those manifested when the patients or animals concerned are not specifically immune to the infection in question.

It is easy enough to show that 'pure line' inbred strains of mice may differ widely in their susceptibility to the same infection. Nearly twenty years ago Sabin found that the standard strain of yellow fever virus used for vaccine production would produce 100 per cent of lethal infections in one strain of mice at a dose which was 10,000 times smaller than that which in a second mouse strain produced no symptoms at all. In this example a single gene difference was involved but such a simple relationship is unusual. Nevertheless it is a general rule that careful experimentation with almost any microorganism that can produce fatal infection will show some degree of inherited difference in susceptibility between almost any two 'pure line' strains of the same animal species.

In man there are apparently well marked racial differences in susceptibility to diseases like tuberculosis, measles and indeed to yellow fever itself as in mice. It is inadmissible however to attribute them simply to genetic differences until environmental factors such as culture patterns and nutritional status are properly assessed. The human situation is so complex that almost the only approach at the genetic level comes from the study of identical twins. Twins can result from the fertilization of two ova liberated at the same time and fertilized each by a different spermatozoon. These are 'fraternal' twins who resemble each other no more than any other members of the same family. Identical twins on the other hand arise when a single fertilized ovum splits into two cells and each of the two gives rise to an embryo with exactly equivalent genetic character. The twins that no one can distinguish apart and who go through school with an absurdly similar record are 'one-egg' identical twins. Much work has been done in recording the medical history of twins, both fraternal and identical, in an attempt to assess the relative importance of inheritance and environmental factors in the origin of disease. There are many difficulties and possibilities of error in such work but with good statistical control and common sense these can be avoided. If the results of a satisfactory investigation show that there is a significantly higher degree of concordance between identical twins than between fraternal twins in the incidence of some type of disease or abnormality, genetic factors must play a part in inducing the condition. The classical use of this approach to show the large genetic element in susceptibility to tuberculosis can be left to Chapter 16 but there is also evidence of a genetic component in determining the likelihood of paralytic poliomyelitis.

Of quite a different character are the rare genetically determined abnormalities of the immune system which render infants exceptionally vulnerable to infection. We have already said something about agamma-globulinaemia, the best known of these deficiency diseases, in relation to its influence on pneumonia and measles. It is known that there are also incomplete and mixed immunological weaknesses of genetic origin which can influence the course of infection. In rather oversimplified form we may say that where the antibody-producing B system is inadequate bacterial infections tend to be severe. Deficiency in the T system interferes with the proper responses to many viruses, to tuberculosis and a number of normally harmless fungal infections. All such conditions are however so rare as to be of no real importance in medicine.

Common sense and analogy can probably allow us to say quite definitely that differences in the reactions of people meeting a given infectious disease for the first time are in part due to genetic differences. We can probably be equally confident that such genetic differences in susceptibility are almost always due to the action of more than one gene. There is however no way of establishing either of these intuitive conclusions and in practice there is nothing to be done about them.

*

The next host factor that calls for discussion is the influence of age on the outcome of first infections. As in so many other contexts recent experience in Western countries has been far from representative of what an ecologist would call the normal behaviour of infectious disease. The widespread use of immunization and the effective treatment of infections at an early stage has changed the whole pattern of childhood illness and mortality. If we want to understand the influence of age it is more enlightening to examine the relevant statistics of infectious disease in Western countries before 1939.

It is important to be clear about the nature of the information we are attempting to study. There are first the statistics of mortality from a given disease. In many ways this is the most useful and least equivocal of all the information we can obtain. The fact of death is incontrovertible, and the diagnosis of a fatal case of infectious disease is, with some qualifications, likely to be relatively accurate. The age of the victim will always be known in any civilized community as well as the size and age distribution of the administrative unit being studied. The incidence of disease, irrespective of whether it is fatal or not, is much more difficult to be sure of. There are several difficulties. Except under unusual circumstances, infectious disease is reported by doctors to the central authority that collates the statistics, on the basis of patients they are called on to visit and on their opinion of what the patient is suffering from. Where, as is so often the case, a given infectious disease can give rise to conditions of widely varying severity, the criterion for reporting will often be very poorly defined. Whether or not mild cases are reported will depend very largely on the interest of doctors or patients at the time. If a disease like smallpox, measles or mumps has well-marked clinical symptoms in the great majority of cases we are on surer ground, but even with these there will be a proportion of patients who do not see a doctor and are never reported. Similarly, during an epidemic of rubella, doctors alert to the possibility that the virus may seriously damage the developing embryo will tend to 'over-diagnose' rubella when many of the fleeting rashes they encounter are actually due to echo or other harmless viruses. If it is known that an epidemic of polio is in progress, many cases of minor feverish upset with headache and stiff neck will be diagnosed and reported as polio which at other times would be lumped with other trivial undiagnosed infections.

Despite the difficulties in regard to the diagnosis of polio which we have just mentioned its changes in age incidence in the pre-immunization years were particularly illuminating. Until 1954 polio was the one important infectious disease of advanced countries which had failed to respond to the general improvement of medicine and standards of living. It had, in fact, increased in incidence and severity *pari passu* with those rising standards. To sort out why it did so is an absorbing exercise in epidemiology, and incidentally one that exemplifies particularly well the way a disease differs in age-incidence according to the circumstances of the population it attacks.

Poliomyelitis was recognized as a distinct disease fairly early in the nineteenth century, but it was always rare and was not considered an infectious disease. The first epidemic of the disease to be described occurred in Sweden in 1887. Thereafter there was an irregular but consistent extension of epidemic poliomyelitis to other countries. In general, the more advanced a country in its standards of hygiene and general living, the earlier did it experience epidemics of poliomyelitis. For thirty years epidemics were almost confined to Scandinavia, north-eastern United States and Canada, Australia and New Zealand. By the 1950s epidemics had become more widely distributed but there was still a general correlation between the prevalence of paralytic polio and high living standards, as evidenced for example by a low infantile mortality. Even in the late 1960s it was evident that while polio notifications had fallen dramatically in all the affluent countries as a result of immunization they were rising in the countries of the developing world in parallel with gradually improving living standards.

The first epidemics in Scandinavia, North America and Australia all showed the typical 'infantile paralysis' picture with the major incidence in children under five. Since about 1920, however, polio epidemics in the advanced countries have always involved older children, especially the five to ten age group, and with the years there has been an increasing proportion of young adults, a group which everywhere shows an abnormally high death rate. In Scandinavian epidemics in 1950–1 the outstanding feature was the large number of adults with severe paralysis requiring respirator treatment.

Some light can be thrown on the meaning of this progressive shift in the age incidence of paralytic poliomyelitis by looking at the so-called 'virgin soil' epidemics which have occurred amongst people in remote island or Arctic communities. Such outbreaks have been observed in Guam, Samoa, New Guinea, St Helena and the Hudson Bay Arctic. In these epidemics the brunt of the disease was borne not by infants and young children but by adolescents and young adults. In the Arctic epidemic of 1949, which was investigated in great detail by Canadian epidemiologists, the small Eskimo community concerned was devastated over its whole age range except that there were no cases of paralysis in infants three years of age or under. The paralysis rate was 40 per cent, the mortality 14 per cent, of the whole population. The only large epidemic of this type which has been accurately reported is one which occurred in St Helena in 1946. A graph showing the age incidence of diagnosed cases in the St Helena epidemic may be contrasted with two comparable graphs, one for the 1942–3 epidemic on Malta and the other for a typical epidemic in the Australian state of Victoria (Fig. 10). These three types of age distribution broadly cover all the epidemics of which we have records.

In many ways the epidemiology of poliomyelitis is the most interesting and illuminating of any infectious disease. To understand it we must first say something about the way infection occurs and how symptoms are produced. We know most about the epidemiology of polio from investigations carried out just before the problem of prevention was

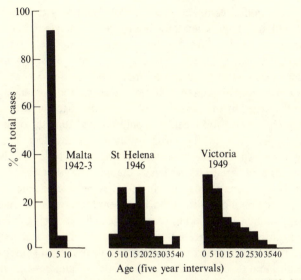

Fig. 10. The age-incidence of poliomyelitis in three epidemics, Malta 1942–3, St Helena 1946 and Victoria 1949. The percentage of the total cases falling in successive five-year age groups is shown.

solved. The situation in the United States around 1950 can therefore be taken as a model. In those days subclinical infection by poliovirus was far commoner than the paralytic disease that was its most important manifestation. In any large American city the virus could usually be shown to be circulating through the community during the late summer months even when there was no frank epidemic. By appropriate tests evidence of infection in the intestine could be obtained from a small percentage of children – but the percentage was large enough to make it likely that almost all children had been infected at least once by the virus by the time they were ten to eleven years of age. Around 1950 there was great interest and some alarm in the fact that in any summer when poliomyelitis was present in New York or Chicago polio virus could be isolated from a few cubic centimetres of city sewage from any of the main out-fall channels. It gives a very vivid realization of the wide extent of invisible infection that went on in the presence of relatively few cases of paralysis. When studies on polio infection were extended to crowded cities of warm climates and low standards of child care – Cairo, Bombay, the native 'locations' in Johannesburg – it became even clearer how widely the virus spreads amongst infants with a minimal number of paralytic cases.

In Western countries infection was most commonly transferred from child to child at the age when they first tended to mingle in play or other activities with children from other families than their own. Close association between children was necessary, but it is still a matter of some controversy as to the common means of transfer. Most epidemiologists would probably agree that unnoticed minor contamination of fingers with excreta is the most important factor, together with a host of other ways by which food can be contaminated directly or indirectly with faeces. Transfer of the virus by air-borne droplets of saliva is probably rare. An episode that strengthened our belief in the 'dirty fingers' theory in Australia was associated with the visit of Queen Elizabeth II to Western Australia in 1954. A polio epidemic was in its early stages and public health authorities were faced with the dilemma of forbidding children from taking part in all public gatherings or running the risk of an explosive flare-up of the epidemic. Their decision was to allow the children to attend the appropriate festivities but insisting that both families and teachers in charge of groups of children should make certain that every pair of hands was washed with soap and water after any visit to the toilet. Many of us were impressed that the incidence of new cases fell during and immediately after the Queen's visit.

The virus enters the body by the mouth, multiplies in the throat or in the lining of the intestine and the body is stimulated to produce the standard immune response. In a week or two multiplication and excretion of the virus ceases and the episode is over without the child or its parents having noticed any ill effect. In less fortunate children the virus passes into the blood circulation, and if it reaches the nervous system, infection will produce effects ranging from simple headache and fever with some changes in the spinal fluid – non-paralytic poliomyelitis – to paralytic disease which in its severest manifestations may be rapidly fatal.

Once infection has occurred the resulting immunity will ensure that if further infectious material is swallowed there is a much smaller likelihood of its becoming implanted on the intestinal wall, and an even smaller one of its producing paralytic infection. This immunity resulting from *subclinical* infection is the first important key to the behaviour of poliomyelitis as a disease. The second key is the influence of age on the likelihood that infection with the virus will result in paralysis. Closely related to this is the third key, the realization that the type of virus concerned in one epidemic may have a higher intrinsic capacity to produce paralysis than another strain of virus. This is independent of the standard types of poliovirus defined by their immunological specificity as

94

I, II and III. Most polio prevalences are of type I and there is clear epidemiological evidence that different type I epidemics differed sharply in their ratios of subclinical to paralytic infection.

In our present context we are concerned only with differences of susceptibility related to age. Perhaps the position can be clarified by a mild exercise in fantasy. If we were to take a whole community of people of all ages, none of whom had ever met the virus of poliomyelitis, and in some way arrange that they were all infected with a single type of virus, we should find that only some of them became paralysed. The highest proportion both of paralysis and deaths would probably be in persons between the ages of fifteen and twenty-five, the smallest would be in infants. If we allow for irregularities in exposure to infection this is very much what is seen in a 'virgin soil' epidemic. By contrast in the crowded community where infection is endemic, primary infection occurs almost always during the first few years of life. Most children are immune by the age of three and virtually all by the time they are five years old. No matter how virulent the current virus all infections, paralytic or non-paralytic, are confined to the under fives, and the pattern of paralysis conforms to the old name, 'infantile paralysis'. The varying incidence, always within that pattern, from year to year is a measure of the virulence, i.e. the ratio of paralytic to subclinical infections, characteristic of the predominant virus strain at the time.

With every improvement in the standard of living, in cleanliness and toilet training in children, and in housing and sanitation, the proportion of children who escape infection in the 'safe' period of infancy increases. They meet their first infection predominantly just before or just after commencing school and the proportion of paralytic cases – the visible part of the epidemic – is correspondingly increased. Special local features crop up in many, perhaps most, epidemics, but there is no escaping the general overall character of the picture. Only the explanation we have given can make it clear why, unlike almost every other infectious disease, poliomyelitis responded to improving standards of comfort and hygiene not by disappearing but by becoming increasingly prevalent.

It must be stressed that the differences we have been discussing concern only the first infection with poliomyelitis. Once first infection is overcome there is a greatly diminished likelihood of the child ever suffering a paralytic infection. This immunity is not necessarily absolute and there are still gaps in our understanding of the relative roles of T and B immune systems in mediating it. Antibody against polio antigens is present as IgA in the intestine and IgG in the blood. It is usual to assess immunity to polio by serum tests which in the 1950s showed that the proportion of children with antibody increased with age. The speed of that increase varied greatly from one place to another. In tropical areas with low standards of living, Guam, Mexico, Egypt and India, for example, antibody against all three types of virus was present in 90 per cent of children by the time they were three years old. In North American cities the process of natural immunization was much slower and more irregular, many children having missed being infected with one or more of the three types of virus.

Such work on natural immunity pointed clearly to the necessity of developing an effective method of vaccination against polio that could safely imitate the dangerous natural process. It is now a matter of history that, first with the Salk vaccine (virus killed with formalin) and later with the Sabin vaccine (live attenuated virus), polio has been effectively banished from all those countries able to make immunization available to all.

So far we have been writing mainly about the better defined diseases with easily recognized symptoms like measles and polio. A glance at mortality statistics however will show that most deaths from infection are ascribed to respiratory diseases including pneumonia and 'influenza' or to the intestinal infections, diarrhoea and dysentery. Although it is possible in most individual cases to provide a relatively complete microbiological diagnosis if the necessary funds and laboratory personnel can be employed, such information is rarely obtained and in any case is irrelevant to the available statistical information. If we now confine discussion to respiratory infections most of the statements made will also apply, with common-sense qualifications, to what happens in the intestinal infections.

Human respiratory infections can be due to a great variety of immunologically distinct viruses and bacteria. No one can become immune to them all and most individual immunities seem to fade within a few years. Further, all the important viruses and bacteria concerned must be relatively highly mutable and capable of giving rise to new immunological types. This is known for the influenza viruses and can be assumed (because of the large number of known types) for the adenoviruses, the various sorts of picornavirus, including over 100 common cold viruses, and the pneumococci of which there are at least eighty types. For practical purposes of interpreting overall statistics we can forget about specific immunity and regard mortality statistics for pneumonia and other respiratory infections as a measure of vulnerability in a general sense.

Death from respiratory infection will in the great majority represent a terminal pneumococcal or other type of bacterial pneumonia, very often initiated by a viral infection. For our purpose the only satisfactory procedure is to pool deaths ascribed to pneumonia, influenza and bronchitis. This has been done in Fig. 11 showing the English experience in two years when influenza was exceptionally severe, the pandemic years of 1891 and 1918, and an intermediate year, 1896, when influenza was absent or insignificant. The graph needs a little explanation before it can be analysed. It is evident to everyone that 'physiological time' is not the same as ordinary time. The rate of growth of a human being is highest early in embryonic life and, with some minor oscillations, slows down progressively till it stops about the age of twenty. One can allow for this in a crude way by using a logarithmic time scale from conception to twenty years and then using the linear scale. In conformity with common epidemiological usage the mortality by ages is shown on a logarithmic scale.

Taking first the years 1891 and 1896, mortality lies along two straight lines which slope down to a minimum at the age of ten to twelve. If there is a real 'prime of life' it must lie in those magic years. It will be seen that the only difference between the influenza year 1891 and the non-influenza year 1896 is in the height of the curves; their slope and conformation are almost identical. Similar curves hold for all the years that have been tested except for those dominated by the influenza pandemic of 1918–19 and a few subsequent years presumably while the same virus was still current. In 1918 we see the appearance of the excessively high mortality in young adults which impressed all observers at the time and which is strongly reminiscent of the age incidence of paralysis in a virgin-soil polio epidemic.

Fig. 11 immediately suggests that there are three phases of human life that are liable to high mortality when infection becomes established in a non-immune individual. It has been known immemorially that infants and the very old were vulnerable; the graph merely shows the regularity with which that vulnerability increases to the two ends of the life scale. The young adult peak is only sometimes visible but it is just as real. A few examples, comments and qualifications about the three vulnerable ages of infancy, young adulthood and old age may be appropriate here.

Mortality during infancy is notoriously susceptible to environmental influences. Adequate nutrition, i.e. breast feeding followed by appropriate diet at the highly vulnerable weaning period, is probably the most significant factor in determining survival, particularly in primitive communities. Another factor that has often been suggested but probably

97

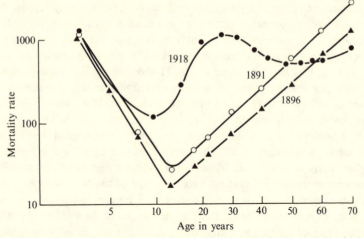

Fig. 11. Mortality rate by age from respiratory disease in England and Wales: (*a*) in a normal year – 1896, (*b*) in an 'ordinary influenza' year – 1891, (*c*) in the pandemic influenza year – 1918. Mortality rate on logarithmic scale; age in childhood on a logarithmic scale changing to a normal scale at twenty years.

never quantitatively established is that there must always be a weeding out by otherwise trivial infections of infants who are genetically inadequate in one way or another.

In any region where infection is prevalent the mother will normally transfer significant amounts of most antibodies across the placenta to the child before birth. Everything suggests that this passive maternal transfer of antibodies provides a relatively effective resistance to the corresponding pathogens for from six months to a year after birth. The evidence for this in man is largely circumstantial but the concept is strongly supported by the well known incidence of 'scours' (diarrhoeal disease) in calves which have not received maternal antibody. In cattle antibody is transferred not across the placenta but by the colostrum from the udder at the beginning of lactation. Human infants rarely contract such generalized infections as measles, mumps or chickenpox during the first six months of life even under conditions of severe exposure. The situation is somewhat different with superficial infections of the respiratory tract because the type of antibody that is relevant to protection of such mucous membranes (IgA) is, unlike IgG, not transmitted across the placenta. For example, one of the currently important problems in paediatrics is infection by respiratory syncytial virus in infants, particularly under six months of age. Here maternal antibody is ineffective – there have even been suggestions that by allowing an antigen–antibody reaction in the delicate bronchioles of the infant lung it may do positive harm.

At the other end of life matters are complicated by the past experience of numberless infections leaving a medley of fresh, faded or vanished immunities. To obtain a clear picture it is necessary to look at the mortality from infections which are so rare in the community that one can be reasonably certain that all diagnosed patients are undergoing their first experience of the pathogen no matter what their age. Two examples are psittacosis, which is hardly ever fatal in persons under forty, and the outbreak of a then new type of mosquito-borne encephalitis in St Louis, USA in 1933 where fatal cases were almost all in elderly people and there was a regular increase in mortality rate with age. In neither was there evidence of a young adult peak.

The third susceptible phase, in adolescence and young adult life, is not commonly seen for the obvious reason that most people in those age groups have developed adequate immunities in childhood. In the other direction apparent examples of such incidence of cases or deaths must be looked at to ensure that the hazard is not simply an occupational one. If hunters and wood-cutters are the only people who enter an environment where infection with scrub typhus or jungle yellow fever is possible, then obviously there will be a concentration of deaths on young adult males. Of the probable genuine examples in addition to polio and 1918–19 influenza, there are indications that a more severe illness occurs at fifteen to twenty than two to five years in smallpox, measles, mumps and chickenpox, and this is seen even more clearly with infectious hepatitis. In the early years of the century tuberculosis deaths, particularly in women, showed a striking peak between the ages of twenty and thirty.

The simplest explanation of the young adult impact has already been briefly stated in discussing the T system immune response. Without elaborating what after all contains a considerable element of speculation, we must regard the T response as always being two-edged in relation to infections. It is responsible for most of the symptoms of measles as well as for ensuring persistent immunity. The effectiveness and vigour of most physiological functions is at a peak in the young adult and this holds in regard to inflammatory reactions of various types. Anyone for instance who has studied skin reactions against viruses in man will know that in children one can obtain a mild response that is specific evidence that the child in question has been infected and, equally important, obtain clear negatives in those who have not. Try the same tests on medical students or servicemen and not only are the reactions much stronger but it is almost impossible to find anyone who is clearly negative even to viruses he can never have met. There will be endless interest for workers in the next decade to work out the full implications of the T system in relation to immunity. For the present we can only express our confidence that the explanation will be found in this direction and end with a few words on the possible evolutionary significance of the findings in regard to age incidence.

No one can doubt that for the last few thousand years most human groups have been living in an environment saturated with infection. An intense selection of those best fitted to resist disease has been constantly in progress. Only in the last century has it been in any way diminished in severity. We can feel reasonably certain that those characteristics we find in human beings now are the best practical compromise for dealing with infection which lay within the species' power to develop. In all probability the pattern evolved was broadly similar to that of other mammals, but ever since urban life developed there must have been a more stringent selection for survival in the presence of a variety of infections than was ever experienced by our pre-human ancestors. If the species was to survive, it was necessary that children should be able to overcome these diseases and develop immunity against them. In a constantly infected environment it was not so necessary that the adults, immune from childhood against the common diseases of their herd, should be relatively insusceptible to diseases they had not previously met. Probably it was more important in the long run that they should possess high resistance to local infections of wounds and abrasions suffered in hunting, fighting and the like.

There is evidence, when we look for it, that human physiology is modified according to age to correspond to these evolutionary requirements. From both angles we reach the same general picture of what may be called the five ages of man. First the infant, needing protection both by the antibodies transmitted from its mother and by a diet which is uncontaminated by bacteria. Then the child, easily infected, but dealing effectively and rapidly with most infections, and during this period building up a basic immunity to common infections which will last through life. The young adult in full physiological vigour is less easily infected and deals rapidly with local infections, but in the absence of a pre-existing immunity is liable to be overwhelmed by his own too vigorous reactions to general infections. Then when the prime of life is past, immunities persist and the unduly vigorous response to generalized infection disappears. Acute infections are much less frequent and there is usually a long period when, barring accident, the individual is free from anything but minor illness. Finally the gradual degeneration of all bodily function is associated with an increasing susceptibility to non-specific respiratory or gastrointestinal infection in old age.

*

The third of the general features which seem to dominate the question of susceptibility and resistance is the region of the body which feels the brunt of the infection. In general when a potentially virulent organism reaches one of the 'sheltered' regions of the body where normally there is no contact with the environment, it is likely to produce a much more serious result than in its natural (or at least usual) habitat.

The best-known example of such susceptibility of sheltered regions of the body is puerperal fever, which results from streptococcal infection of the uterus immediately after childbirth. The infection is severe, and in the days before the sulpha drugs and penicillin about a quarter of the cases ended fatally. The uterus after childbirth is in rather a disorganized state, its stage of active function is over, and it has not yet commenced the process of shrinking and reorganization which will return it to the normal non-pregnant form. There is always at this stage a certain amount of blood clot and dead tissue cells to provide food for any bacteria which may enter the uterus, and the blood supply of the organ is in a makeshift, intermediate stage, poorly adapted to deal with the emergency of an infection. If a sufficiently virulent *Streptococcus* does reach the uterus, a very serious infection results. There are effective bacteriological techniques for tracing the origin of such streptococcal infections as puerperal fever, and it is common to find that they are derived from nurses or other attendants with a mild streptococcal sore throat.

Another interesting example of the way a relatively harmless bacterium can be responsible for fatal infections if it happens to reach the wrong place, is given by the Bundaberg tragedy of 1928. Bundaberg is a town in Queensland, Australia, where, in the summer of 1928, an immunization campaign against diphtheria was in progress using toxin–antitoxin mixture, at that time the standard immunizing agent. By a chain of unfortunate circumstances, a bottle of the inoculation mixture became infected with a *Staphylococcus*, presumably picked up by the needle from the skin of one of the inoculated children. The bottle was left in a cupboard in warm subtropical climate for a week and then used for inoculating twenty-one young children. Unknown to the doctor, the material now contained millions of staphylococci which were injected under the skin of the children. The result was disastrous; twelve of the children, particularly the youngest ones, died within forty-eight hours, some of the others were severely ill but recovered, while four showed practically no symptoms beyond the formation of a small abscess at the point of inoculation.

The tragedy of course created intense public interest, and a Royal Commission was immediately appointed to investigate the circumstances. It was at once discovered that the bottle of toxin-antitoxin mixture was heavily infected with staphylococci, and that similar bacteria were present in the local abscesses which formed at the site of inoculation in those children who survived. The obvious inference was that the staphylococci had caused the deaths, but the extraordinary rapidity and severity of the symptoms was quite unlike the effect ordinarily produced by staphylococcal infections. After all, the average boil on the back of the neck can be painful but it is more a nuisance than a danger. In its core there are probably many more staphylococci than were injected into any of the Bundaberg children, but the essential difference is that those numbers are built up gradually and by the time they are significant the staphylococci are surrounded by a thick defensive screen of phagocytes and their auxiliaries. But at Bundaberg something totally unnatural occurred. It is biologically unprecedented for several hundred million staphylococci to appear suddenly in healthy normal tissue underneath the skin. In the twelve children who died and the five who became dangerously ill this unnatural invasion overwhelmed the local defences, and the bacteria multiplied and spread throughout the body. The other four children, however, were hardly affected at all; they had no symptoms until two or three days after the inoculations. Then they noticed a tender lump where the injection had been made and within the next few days saw a small abscess develop here. These children had a local defence adequate to deal even with this invasion. It is probably significant that most of them were older children who might be expected to have had more 'training' in dealing with staphylococci. Almost certainly they had in previous years developed enough immunity against staphylococci to render their phagocytic response to the invasion vastly more effective than it was in the unfortunate non-immune children.

While we are discussing the differences amongst body tissues in their reactions to infection, we should mention two special regions. The first is the central nervous system, the brain and the spinal cord. As the most delicate and important part of the body, this is very carefully shielded and insulated from the environment physically and physiologically. There is even a special physiological arrangement which prevents soluble substances in the blood from entering the actual substance of the nervous tissue. It is very rare for infection to lodge in the brain, even if the organisms are passing through it in the blood. As a natural corollary to the existence of this elaborate shielding of the nervous system, if bacteria do enter it by any means, they are very liable to multiply and cause rapidly fatal illness. Meningitis, which is an infection of the surface coverings of the brain and spinal cord, is only rarely caused by the common streptococci, staphylococci and pneumococci, but when such infection does occur, it demands strenuous treatment with antibiotics if the patient is to survive. The bacteria that cause meningitis in childhood are chiefly two relatively harmless denizens of the human throat. The *Meningococcus* is the more important of the two. This bacterium is rather misnamed. A very minute proportion of the infected individuals develop any signs of meningitis. Normally it is a harmless inhabitant of the throat, but it has a distinctly greater aptitude for reaching and infecting the brain surface than any other of the organisms living in the throat. The other fairly common cause of meningitis in infancy is the bacterium, *Haemophilus influenzae,* another inhabitant of the throat and air passages, and like the *Meningococcus* practically never producing any serious results there.

The other tissue which differs sharply from the rest of the body in its response to infection is the special type of bloodless fibrous tissue that forms the valves of the heart. These thin strong cusps of rather parchment-like appearance are continuously surrounded by blood, and there is no need for blood vessels within them to supply nourishment. Sometimes, especially when the valves are congenitally defective or damaged as a result of an earlier attack of rheumatic fever, some types of streptococci become implanted on a valve. They remain there, growing slowly but persistently, and gradually destroying the valve. The common cause of this disease, subacute bacterial endocarditis, is a type of *Streptococcus* very common in the mouth, and often associated with minor dental disease such as apical abscess. The commonest reason for its entry into the blood is the extraction of an infected tooth. In a normal person this little flush of streptococci into the blood does no harm, but in a person with a heart damaged by rheumatic fever it is potentially very dangerous. It is general practice to remove teeth from such a person only under the protection of a short course of penicillin injections. In any other tissue than the heart valve, the presence of the *Streptococcus* would provoke all the usual defence responses of blood vessels and phagocytes. But on the valve, although its blood supply in one sense is unlimited in amount, yet it is not suited to deal with unwanted bacteria. Red cells and white cells rush past at a speed comparable to that at which water flows from a bath tap. There are no capillaries through which the blood trickles gently, allowing phagocytes to stick to the walls and pass into the tissues where they are needed. So we have the curious anomaly of a relatively harmless bacterium producing steady and ultimately fatal damage in a situation where it is exposed to the full force of the blood stream, but remains unharmed by the agents in the blood which anywhere else would soon destroy it.

8. How infections spread

If we are to make any attempt to prevent or check epidemics of infectious disease, it is obvious that the most helpful thing we can know is how the infection spreads from person to person. Just to know that an infection is spread by contaminated water or milk or, to take another example, by mosquito bite, will often be sufficient alone to allow adequate public health measures to be taken. Typhoid fever, for example, was being rapidly eliminated from England by attention to sewerage and water supply well before the typhoid bacillus was discovered. Even more striking is the story of yellow fever. In 1900 the American Yellow Fever Commission proved that the disease was mosquito-borne, and on that basis yellow fever was almost wholly stamped out from Central America. But the virus responsible for the disease was not isolated until 1930.

In discussing the spread of infection, we ought logically to consider first the way in which the infective agent is liberated from the originally infected person or animal; secondly, how it passes from the infector to the new susceptible host; and thirdly, how it enters the tissues of the latter and provokes infection. In particular diseases, any of these aspects may be of great importance, but in practice we can divide infections into four main groups on the basis of the way infection is transferred.

Our first group comprises those infectious diseases that infect primarily the digestive tract, usually concentrating on the lower part of the bowel. These are spread by the dissemination of faecal material containing the responsible microorganisms. Water, milk, food or eating utensils may be contaminated in various direct or indirect fashions. Fingers and flies are probably the two most potent agents for indirect transfer, but however devious the route from the source of infection, the microorganisms are eventually swallowed, and passing through the stomach infect the bowel. Apart from amoebic dysentery, due to a protozoon, the best-known infections of this group are of bacterial origin. They include typhoid and paratyphoid fever, cholera, dysentery due to a wide variety of bacteria, summer diarrhoea of infants and bacterial food poisoning.

For many years it was thought that there were no virus diseases of the intestinal tract. In a sense that is still true, though in children it is common to find short episodes of diarrhoea which are associated with virus in the intestine and probably caused by it. On the other hand, there are about a hundred distinct viruses of echo, coxsackie and adenovirus groups that can multiply in the intestinal wall usually without producing any symptoms. Occasionally one of these viruses will spread through the bloodstream to involve other organs such as brain, skin or muscles. A variety of symptoms may result – fever, rash, headache and neck rigidity or sore throat and painful cramps of the respiratory muscles. Polio virus is a typical member of these 'enteroviruses', unique only in its capacity to paralyse. Another virus with no close resemblance but equally spread by the intestinal route is that of infectious hepatitis.

The method of preventing such diseases is obvious to everyone. Indeed, the one indubitable blessing of modern civilized life is the development of the technical methods and mental attitude required to eliminate them. That has been well done, and can be justly claimed as the best contribution the English-speaking peoples have made to humanity. Decent sewage disposal, pure water supply, pure food laws, control of milk supply and pasteurization, plus the cult of personal cleanliness have rendered most of these diseases rare in any civilized community. An outbreak of typhoid fever or a high infantile mortality from diarrhoeal disease is rightly regarded as a civic disgrace.

It is interesting to look back on the development of such a civic conscience and to realize that its all-important beginnings were based on a completely wrong idea of how infections like typhoid fever arose. In the early nineteenth century it was recognized that typhoid fever and filthy drains went together, but the stress was laid more on the smell than on possible infective principles (bacteria were, of course, unknown) in the objectionable drainage. By some transference of ideas, bad-smelling drains were also blamed for the incidence of diphtheria, and the incentive to remove these two diseases was largely responsible for the development of modern sanitary engineering. The process was well under way by the time the infectious nature of typhoid fever was clearly recognized, and it was a good deal later that the typhoid bacillus was found. Both these discoveries, of course, accentuated the necessity for keeping drinking water and sewage, to put it bluntly, out of each other's way.

For many years it has been the standard practice to protect against typhoid fever by inoculation with a vaccine, usually the TAB vaccine which is designed to prevent paratyphoid fevers A and B as well. Antityphoid vaccination seemed to work very well during the two World Wars but, of course, the medical authorities of all armies also insisted on the best possible field hygiene. It is a valuable commentary on how difficult it is to evaluate the role of a given procedure in the presence of other possibly relevant factors when we have to confess that most critical bacteriologists still have doubts about the protective capacity of typhoid vaccination. It still remains a routine precaution and will go on for a long time no doubt but 'environmental sanitation' is the only measure that can be relied on to eliminate typhoid fever.

There are still some riddles to be answered about intestinal infections. For some diseases the results of conventional improvements in sanitation seem better than we could reasonably expect. The more serious diseases, typhoid, cholera and the more severe forms of dysentery, have been almost eliminated but lesser forms of gastrointestinal infection, notably the type of dysentery caused by *Shigella sonnei,* continue to spread readily. Polio and infectious hepatitis are just as much intestinal infections as typhoid fever or dysentery yet they can spread easily under conditions when the bacterial diseases do not occur. As recently as the 1950s most urban children in the United States had been infected (subclinically) with at least one type of poliovirus by the age of ten. Since the disappearance of polio as a result of immunization, infectious hepatitis has become the most important disease of childhood and adolescence. Clearly our current methods of handling sewage are not adequate to prevent the continued circulation of the causative viruses in the modern Western city and most epidemiologists feel that an effective vaccine against infectious hepatitis is almost the greatest public health need in 1971. A minor unsolved problem in the same field is the unholy capacity of threadworms to spread by transfer of eggs from one child to another. Much to the embarrassment of hygiene-conscious households threadworms are still common in the very best of families!

The simplest and most likely explanation of this apparent anomaly is that it depends on whether faecally contaminated fingers can transfer enough organisms to comprise an 'infectious dose' of that particular agent. For instance, hepatitis virus particles may well be excreted in larger numbers than typhoid bacilli or more typhoid bacilli than hepatitis particles may be needed to initiate a typical infection.

The next method by which infection can spread is by far the commonest and most important in civilized countries, namely the respiratory route, or 'droplet infection' as it is usually called in technical accounts. Suppose we have a cold at the early stage when there is a steady call on the handkerchief. If the fluid leaking from the nose contains one of the hundred viruses responsible for a common cold, then there is also not the slightest doubt that the virus is also present in the saliva. When we sneeze, cough or shout, droplets of saliva are forcibly expelled into the air. Visible droplets settle rapidly to the ground and rarely convey infection. It is the very tiny droplets which are important. Unless the air is saturated with moisture, these droplets almost immediately evaporate. They are not much larger than the droplets which makes one's breath visible on a frosty morning and evaporate almost as rapidly. The frosty morning's breath is due to condensation of water vapour to tiny spheres of water which can again be dissipated completely into vapour, but the droplets of saliva contain other things than water. When they evaporate they leave as a residue a tiny light flake of dried protein in which are held any viruses or bacteria which were present in the droplet. Such flakes take a long time to settle in still air, and are easily carried about on air currents. Probably with every breath we take in a room where there are more than one or two people, some of these flakes from other people's saliva pass into our noses. There is an efficient method for filtering most of them out of the air before it reaches the lungs and when we get down to actual figures of infection as judged either by symptoms or antibody rises, we find that even with a virus of the high contagiousness of influenza, only a small proportion of contacts contract the infection. During the Asian influenza epidemic in Melbourne in 1957 it was calculated that each infected individual on the average transferred the virus effectively to no more than two other persons. Sooner or later with an increasing number of immune individuals this rate of spread becomes inadequate to maintain an epidemic and the outbreak then terminates.

There are several different areas where infective droplets taken in by nose or mouth can lodge and if conditions are right initiate infection. Colds seem to start sometimes in the lining of the nose, sometimes in the larynx and then spread up or down; influenza probably results from virus particles lodging in the bronchial passages. Many viruses that produce mild upper respiratory infections ('colds') in adults may be more dangerous to infants, producing serious infection of the lungs manifested as bronchiolitis or pneumonia. Diphtheria and the various forms of streptococcal sore throat have a predilection for the tonsils, while the agent of psittacosis seems to be harmless unless it can be carried right down into the lungs. These differences may be reflected in the alternative ways by which infections of this group can spread.

Before the days of universal pasteurization, streptococcal sore throats, including scarlet fever, were rather frequently a result of milk infection, sometimes resulting from contamination of the milk by infected dairy employees, sometimes actually from a streptococcal infection of the cows. Since all food and drink passes the tonsils on its downward journey, it is only to be expected that those organisms which tend to infect that region may be spread as readily in milk as by the droplet method. Psittacosis, on the other hand, has to reach the depths of the lungs, and to do so the organism must be in very finely distributed form and probably in rather large amount. Dust from infected parrots is the only common source of infection which fulfils these requirements. In one outbreak in Louisiana, probably derived in the first instance from egrets, there was a sequence of infections from patient to sick-bed attendants presumably associated with the intense virulence of the strain of organism concerned. In general, person to person infection does not occur with psittacosis and there is nowhere any suggestion that the infection can be transferred by food or drink.

It has been calculated that about half of all episodes of human illness are caused by respiratory viruses. Most of these are of course quite trivial infections like the most frequent of them all, the common cold. Add to this the fact that most of the generalized infections of childhood such as measles, chickenpox, mumps and rubella are also spread by the respiratory route, and it becomes clear that, in advanced countries, droplet infection is much the most important route by which infectious diseases spread. Whenever any exceptionally serious outbreak of one of these diseases occurs, pandemic influenza for example, efforts are always made to check its spread. Quarantine and isolation restrictions are enforced, and sometimes the wearing of gauze masks over the mouth and nose has been required. It is doubtful whether such measures are of the slightest real value unless the occasion is serious enough to justify complete dislocation of community life. When a few cases of pneumonic plague appear, public opinion will allow and demand complete segregation of the infected persons and their contacts and enforce it with machine guns if necessary. Those who have to treat the patients will adopt rigid precautions which could only be applied in a well-equipped hospital. They will wear gowns and rubber gloves which can be sterilized after each visit to the patients, and protect eyes, nose and mouth with goggles and gauze masks. Modern techniques of hospital air-conditioning and sterilization of air by filtration or ultraviolet irradiation can also be applied to protect other patients and staff from the dangerous 'shedder' of organisms. This approach is the direct converse of the commoner modern situation where a highly vulnerable patient with extensive burns or on immunosuppressive drug treatment must be protected from air-borne microorganisms from other patients and staff. In the former case the patient will be placed in a room under negative

109

pressure and the air sterilized on exit; in the latter the patient's room will be under positive pressure and the air sterilized on entry. In one way or another the spread of droplet infection can be stopped if stopping it is the full-time job of all concerned, but not if persons in the infected community have to go on with their daily business.

It is a clear enough indication of this that while the enteric infections have declined sharply under the impact of modern hygiene, respiratory infections are as common as ever – perhaps more so in the industrial cities of the Western world.

The venereal diseases exemplify a third method of spread involving close bodily contact. There are five human diseases in this group, all caused by bacteria, some of them very odd ones. They are: syphilis, due to a corkscrew-shaped spirochaete; two due to readily cultivated bacteria, gonorrhoea and soft chancre; a chlamydial disease (lymphogranuloma venereum) distantly related to psittacosis and trachoma; and a tropical disease due to an unusual bacterium (*Donovania*). All these micro-organisms are very readily killed outside of the body, and very direct contact is necessary for their transmission. Both of the common forms, gonorrhoea and syphilis, are readily cured by penicillin, and, more important from the point of view of prevention, the patient is rapidly rendered non-infectious for others. Lymphogranuloma venereum is not susceptible to penicillin but can be cured by some of the newer antibiotics.

During the war venereal disease was an important military problem, and most of the measures which have been suggested were tried in various combinations. These included direct disciplinary measures, the provision of opportunities for sport and other leisure-time employment, moral and religious exhortation on the one hand, the issue of pro-phylactic packages to men on leave and the establishment of depots for disinfection after exposure on the other. These measures probably had a considerable effect, but they did not prevent the general rise in the incidence of venereal disease that has characterized every war in recent history.

With the post-war return to normal social conditions the incidence of venereal disease gradually diminished and maintained a relatively low level in all Western countries. There was even hope that syphilis might be a vanishing disease. Unfortunately in 1971 we have to report that syphilis and gonorrhoea are rampant in almost all Western countries. The improvement after 1947 could be ascribed to the effectiveness of penicillin treatment and the reintroduction of normal social life after the war. The epidemic spread that we are now experiencing is blamed by many authorities on the contraceptive pill. This has produced a greatly increased freedom of casual intercourse in young people, freed from the threat of pregnancy and confident that penicillin can deal with any venereal infection. The medical implications of this new sexual revolution are not easy to predict.

There are types of personal contact other than venereal which can convey other types of infection. Kissing can infect a susceptible partner with herpes simplex (cold sores) or with infectious mononucleosis (glandular fever). The widespread and important infection of the eye and eyelids, trachoma, may be spread by flies but is probably more often conveyed directly to the eye by contaminated fingers, towels, etc. Trachoma is particularly common in the hot dry and sandy regions of Asia and Africa where standards of hygiene are still primitive. The importance of this disease can be gauged by the fact that 20 million of the world's blind are victims of trachoma and nearly half a *billion* people are chronically infected.

While on the subject of contact infections it is appropriate to mention certain diseases that are not spread directly from man to man but do enter the body by way of the skin. The best known are the tropical infections with helminths such as hookworm, but a particularly interesting instance is leptospirosis, the classical form of which is also called Weil's disease. Leptospirae are small corkscrew-type organisms which in man may produce fevers ranging from one with severe jaundice and often fatal, to a mild three-day fever that can only be diagnosed by appropriate laboratory tests. The organisms are natural parasites of rats, pigs, dogs and some wild rodents, and are excreted in the urine of infected animals. They can survive for relatively long periods in polluted water. These leptospiroses are characteristically occupational diseases. The sewer worker or the man cleaning up a fish market is liable to contract the severe form that comes from infected rats. Some years ago there were extensive outbreaks amongst Queensland sugarcane cutters working in wet fields that had been contaminated by a native species of rat with a special type of *Leptospira*. Another milder human disease common in Australia and central Europe where it is called *maladie des porchers* (swineherds' disease) is due to yet another *Leptospira*, named after a little orchard town, Pomona, in southern Queensland where it was first observed. Pigs and cattle provide the reservoir and again the organism is liberated in their urine. All these infections result from skin contact with contaminated water or mud, the organism penetrating not through intact skin but through cracks and minor wounds.

Natural history of infectious disease

In the three methods so far discussed by which infection is transferred, the organism enters the body by one of the natural routes. The last important method involves the 'unnatural' introduction of infection into the blood or tissue by the bites of animals, usually, but not always, insects or ticks. In this group we have the complication that the transmission of infection is frequently not from man to man, but from some infected animal to man. There are now no common human diseases spread by this method in temperate climates amongst civilized communities, but malaria, plague and typhus were once widespread in England and Europe generally, and would probably return with any breakdown of civilization. All these infections have their own peculiar characteristics, which are largely determined by the habits of the animal that transmits them. It is not easy to discuss them in general terms, and the best introduction is perhaps to give a table showing the more important diseases of the group, the arthropod responsible for transmitting the disease to man (technically the 'vector') and the species from which the vector derives the infection. This species is referred to as the 'reservoir' of infection.

Among the human diseases of this group we find an interesting range from diseases which, biologically speaking, have nothing to do with man to those for which he is the only susceptible species. At the former end of the scale we have such conditions as Rocky Mountain spotted fever or the scrub typhus of the eastern tropics. These diseases are due to rickettsial microorganisms which are in nature relatively harmless parasites of wild rodents transmitted from one to the other by ticks or mites. It is a pure accident that man happens to be highly susceptible to the infections. Then we have those parasites that are partially adapted to the human host, such as those of typhus fever and African sleeping sickness, and finally the diseases that are specifically human, such as malaria, dengue fever and urban yellow fever.

Malaria is the most important of the insect-borne protozoal diseases and yellow fever of the similarly transmitted viral diseases. A separate chapter will be devoted to each of these diseases to allow discussion of their complex ecology. Other factors that are important in the transfer of disease from animal to man are better considered in Chapter 10. It is appropriate however to say something here about the process by which insects, ticks and the like convey infection to man. Sometimes a biting insect may act as a purely mechanical carrier. Myxomatosis of rabbits is spread in Australia by any mosquito that will feed on rabbits. An infected animal has a large amount of virus in the skin swellings characteristic of the disease, which for obvious reasons occur on the

Table 2. *Important human infections derived from animals*

Disease	Microorganism	Vector	Reservoir
Malaria	Protozoon	Mosquito	Man (monkey?)
Sleeping sickness	Protozoon	Tsetse fly	Man, mammals
Kala azar	Protozoon	Sandfly	Man, dog
Plague	Bacterium	Flea	Rat, other rodents
Relapsing fever	Spirochaete	Tick, louse	Rodents, man
Yellow fever	Virus	Mosquito	Man, monkey
Dengue	Virus	Mosquito	Man
Encephalitis	Virus	Mosquito	Birds, mammals
Haemorrhagic fever	Virus	Mosquito, tick	Mammals
Typhus I	Rickettsia	Louse	Man
Typhus II	Rickettsia	Flea	Rodents
Scrub typhus	Rickettsia	Mite	Rodents

regions where mosquitoes are most likely to feed. As its stylus probes into blood capillaries, the insect mechanically contaminates its mouth parts with virus. Some of this is left behind when the next susceptible rabbit is bitten. The situation can be exactly imitated by pricking a pin into a myxoma lesion on one rabbit and an hour or two later making a gentle prick with the same pin on the nose or ear of a normal rabbit, hence it has been called the 'flying pin' method of spread.

More frequently insect transfer shows the marks of a long evolutionary history. We have already said something about the evolution of some of the protozoa, including those responsible for sleeping sickness and malaria. The arboviruses, which include yellow fever virus and over 200 similar viruses spread by mosquitoes in tropical and subtropical regions, show remarkably similar behaviour. The virus is taken into the mosquito's intestinal tract with a blood meal from an animal or bird in its first few days of infection. Infection of insect cells occurs and the virus passes first into the fluids of the body cavity and after eight or ten days into the salivary glands of the mosquito. As soon as this happens the mosquito is infectious for any animal, bird or man that it can bite.

A third type of infection is seen particularly in louse-borne typhus fever, due to a rickettsia. Here the louse takes in the rickettsial organisms in blood and the cells lining its intestinal tract are invaded. Soon large numbers of rickettsiae are being voided with the faeces. Louse bites itch, and to scratch them is almost inevitably to work living rickettsiae into the puncture wound made by the louse or the superficial abrasions from scratching. The usual mode of infection with rickettsiae transmitted by fleas (murine typhus), ticks (Rocky Mountain spotted fever and others) and mites (scrub typhus) is probably of similar character. However, highly infectious faeces are also liable to be infectious by inhalation in dust.

Before leaving the topic of the transmission of disease by arthropod vectors we should mention the fact that some pathogenic organisms can pass from tick to tick by passage through the egg. The best known example is in tick-borne relapsing fever where transovarial transmission of the spirochaete through successive generations of ticks has been observed. The rickettsial disease, scrub typhus, is transmitted by a mite which in its normal cycle of development takes only one blood meal in the larval stage. Transovarial passage of the rickettsia in the mite is therefore a necessary phase in the process of human infection.

One could almost make the general statement that if there is any conceivable way by which a disease-producing microorganism in one species can be transferred to another susceptible species, some organism will be found to be making use of it. The story of the plague flea's regurgitation of the bacillus from his blocked oesophagus can wait until Chapter 17 but perhaps the story of human rabies in South America from the bite of vampire bats is even more exotic. Since 1921, there had been recognized epidemics of paralysis amongst South American cattle. These became gradually more frequent, and in 1928 there was a very extensive outbreak which caused the loss of something like 30 per cent of the total cattle in Paraguay. The disease was found to be a form of rabies (or hydrophobia), in which the violent 'mad dog' type of symptoms were replaced by paralytic ones. There is little rabies in dogs in South America, certainly not nearly sufficient to account for such widespread destruction amongst cattle. Gradually suspicion lighted on the vampire bat, though it was known that these bats had taken blood from cattle for centuries without causing any serious harm. In 1930 a further development was observed on the island of Trinidad, thousands of miles north of Paraguay. Here a similar disease broke out amongst cattle, but in addition about a dozen human beings contracted the same fatal paralytic disease. Again vampire bats were incriminated, and the crime was sheeted home by the discovery of several bats which actually harboured the virus. Vampire bats gain their meal by stealth, usually it is said from the big toe of a sleeping victim. Their incisor teeth are razor-sharp and can excise a little hole in the skin without disturbing the sleeper and lap up the blood as it seeps into the hole. The virus is in the saliva of the bat as it is in that of a rabid dog, and if the bite is by an infected vampire, transmission to the bitten individual is almost certain. Several months may elapse before symptoms appear in the victim but eventual death is almost inevitable. Rabies is the most lethal of all infectious diseases. As far as one can gather, the disease spreads from bat to bat in their roosting places, not by way of infected cattle or other victims. Some of the infected bats develop symptoms and die, some remain healthy carriers of the virus for months. It is not difficult to protect human beings, who are bitten only during sleep, by suitable wire screening of houses, but protection of cattle is a more difficult problem. In 1956 a new danger emerged with the recognition that rabies could be spread at times by ordinary insect-eating bats. This has introduced a new worry to the epidemiologist, especially as there is recent evidence that such bats could represent a basic reservoir of infection. Of more immediate importance is the report of one or two fatal cases of rabies in speleologists who had apparently inhaled droplets of infective saliva from bats haunting caves they were exploring.

Finally we must not forget the existence of what is known as 'vertical transmission', in which infection is transmitted from one generation to the next, usually by passage across the placenta from mother to the foetus *in utero*. There are two distinct types. In the first, transfer to the foetus is a mere accidental complication of the maternal infection and plays no part in the normal process by which the microorganismal species survives. Congenital syphilis has been known for centuries, but in recent years the capacity of a comparatively trivial virus, rubella, to produce congenital deformity in infants has provoked much more attention. The discovery that a baby could be born blind or deaf when the mother had rubella early in her pregnancy was due to an Australian ophthalmologist, Gregg. He reported in 1940 that he had seen a large number of cases of congenital cataract (opacity of the lens of the eye) in babies whose mothers had suffered from german measles (rubella) during their pregnancy. It rapidly became clear that, in addition to these eye changes, other babies were being born deaf or with congenital malformations of the heart as a result of infection in the uterus during the early months of their mother's pregnancy. In Australia during the years 1939–42 there were probably 400 to 500 infants born with congenital damage of one sort or another resulting from rubella infection. Elsewhere in the world sufficient similar cases were discovered to make it clear that the Australian experience was not unique, but not until 1964 was a similarly large prevalence of rubella in America shown to be accompanied by an equivalent toll of prenatal damage.

For a number of years there was a persistent question as to whether this capacity to produce antenatal infection was a new property of the rubella virus first manifested in the Australian wartime epidemic. Why otherwise had so common a disease as german measles never shown this frightening ability to invade the embryo before? A wholly convincing answer was obtained by the simple procedure of tabulating the birth dates of all persons listed as deaf-mutes on census and other records. It was found that in Australia there have been several periods where many such birth dates are concentrated within a few months. In 1900, for instance, the graph of births of people subsequently registered as deaf-mutes provided a typical epidemic curve whose peak was six or seven months after that of a known rubella epidemic. The answer then is simply that congenital damage by rubella in pregnancy had been occurring for at least forty years before Gregg recognized it in 1940. There had been no change in the virus. It was simply that nobody had been interested enough to wonder why so many deaf-mutes seemed to have been born about the same time.

Other infections may produce damage to the foetus, particularly those whose symptoms in the mother are too insignificant to be noticed. A protozoon, *Toxoplasma,* is known to be responsible for a condition characterized by mental deficiency and eye abnormalities, a few examples of which can be found in any institution for mentally defective children. In adults, who may be infected from contact with a variety of domestic or wild animals, symptoms are usually negligible, though sometimes a condition resembling glandular fever may develop.

This leads us to another agent, the cytomegalovirus, which occasionally crosses the placenta to produce a syndrome very similar indeed to toxoplasmosis. A high proportion of mentally defective children born with microcephaly (a small head and brain) are the innocent victims of transplacental cytomegalovirus infection. Ironically, the mothers are completely without symptoms, for they are chronic carriers of the virus which is usually acquired by salivary spread and persists harmlessly in the salivary glands for years on end.

A somewhat different case is that of a related herpesvirus, herpes simplex. This virus rarely crosses the placenta, but when it does, causes a lethal infection of the baby. Routinely however, it is transmitted by another route which could be loosely described as 'vertical spread',

116

namely via saliva exchanged in kissing of the infant by indulgent parents or in many other ways. Herpes simplex infections in man have some interesting evolutionary antecedents that are discussed at the end of Chapter 10. Serum hepatitis is much less clearly understood but the virus may be transmissible from mother to foetus *in utero*; in view of its importance in several other contexts we leave its discussion to Chapter 20.

9. Epidemics and prevalences

It is conventional to divide infectious diseases into those that are 'endemic', i.e. always present to some extent in the community, and 'epidemic', i.e. showing a sharp concentration of cases in time and space. Typhoid fever is not endemic in Melbourne or London because effective sanitation and good water supply allow no opportunity for extensive spread, but it is endemic in most tropical cities where such facilities are inadequate. Nevertheless occasional epidemics of typhoid occur in Australia – or India – when a deterioration in the *status quo* leads to a substantial outbreak of the disease.

We find this is a very unsatisfactory dichotomy which is not really applicable to the diversity of the situations that occur in any actual community. It may be more useful and more realistic to outline the important factors that mould the nature of prevalence of infectious disease and to illustrate their action by examples, many of which will be drawn from Australian experience.

In the first place it is self-evident that if the microorganism responsible is not present in a community then no cases of that disease will occur. It is one of the great natural advantages of Australia that as an island nation remote from the continental sources of most infectious disease it is relatively easy to prevent entry of disease. In 1971 there are none of the major plague diseases of man or domestic animals in Australia. Nevertheless here, as in all other countries, there are very large numbers of species and varieties of potentially pathogenic microorganism present in the human communities and in domestic animals. These are either not important enough to justify the effort and expense of trying to eradicate them or experience here and elsewhere in the world has indicated that the effort would be unsuccessful. The epidemiological character of the diseases these organisms produce will depend on a number of factors.

First, on the qualities of the microorganism that influence its interactions with the susceptible human host. The route by which infection can occur and the dose (number) of individual organisms needed to initiate symptomatic infection are both important, while an even more significant feature is the ratio of clinically visible infections to invisible but immunizing (subclinical) infections.

Secondly, on the number of organisms shed by the patient or carrier of infection and the period over which such liberation persists, the physical form in which the organisms are shed into the environment and how long the material will remain infectious under existing environmental circumstances.

Thirdly, on the probability of contact with a new susceptible host within the period that the patient or carrier is infectious.

A fourth factor is strictly speaking implicit in the third but must be spelt out separately. This is the extent to which the existing population is specifically immune to the disease we are considering, either as a result of natural infection in the past (clinical or subclinical) or from artificial immunization.

If these four factors can be assessed with reasonable accuracy most of the epidemiological behaviour of the better-known diseases is easy enough to understand. However, nothing biological (outside of a biochemical laboratory) is completely predictable and when we come to deal with actual occurrences of infectious disease we must always be prepared for surprises.

Central to any understanding of epidemiology is some appreciation of the nature, frequency and importance of subclinical infection. It is simplest to define subclinical (or inapparent) infection as the situation where an individual can be shown to be infected, either by isolation of the organism or by the retrospective recognition of an immune response (by skin test or by noting the appearance of antibody in the blood), yet shows either no symptoms at all or none typical of the disease.

Perhaps the most important finding about infectious disease in recent years has been the recognition of the immense variety of trivial infections that are always current in a large city. In the pre-antibiotic era many investigations showed how frequent were invisible epidemics of streptococcal infections of the throat, particularly in orphanages and similar closed institutions of young people. Subsequently, epidemiological study of the spread of antibiotic-resistant staphylococci indicated that these bacteria too are carried asymptomatically, usually in the nose or on the skin, and pass readily from one person to another.

The virologists, however, have found a much richer field of trivial infection in childhood. An orphanage outside Washington, D.C., had an average population of fifty young children with a rather rapid turnover which ensured the entry into the institution of most of the viruses circulating in the community. For a year each inmate was tested weekly for virus infection and more frequently whenever suspicious symptoms

119

appeared. The results showed almost a continuous sequence of 'epidemics', most of them silent but with a proportion of respiratory or intestinal symptoms accompanying some. One of the most interesting was the spread of poliovirus type 3 to involve the majority of the current inmates without a sign of paralytic polio or any other of the standard symptoms in a single child. Every parent knows of the short-lasting minor fevers and upsets of young children. Recovery is complete in a day or two and except under very special conditions no real diagnosis is ever made. These episodes can be looked on as the occasional emergence into symptomatic illness of the wider process by which the child is learning to live with the various microorganisms in the environment. It is something that must have gone on in similar fashion ever since man began to live in cities. In a well adjusted host–parasite relationship, subclinical infection is the rule, disease the exception, and death a rarity.

Our first factor, the specific qualities of the interaction between host and parasite, has two main aspects: the size of the dose, i.e. the average number of organisms required to infect a susceptible individual, and the standard route or method by which infection enters the body. This may range from droplet infection by inhalation into nasal passages or lungs, to wound infection in a surgical ward or inadvertent infection by the doctor's or nurse's syringe.

Even in the healthy individual different microorganisms differ greatly in the number required to produce symptomatic infection. Every professional bacteriologist knows that if he uses normal precautions he will be very unlucky if he suffers a laboratory infection from typhoid, diphtheria or tubercle bacilli. Yet despite the use of exactly the same precautions if he works with the organisms responsible for Q fever, Malta fever or tularaemia he is almost certain to become infected before long. If we had enough information we could probably express the difference in quantitative form: that if one living organism of typhoid is taken into the body the chance of its initiating infection is one in a thousand but if a single Q fever rickettsia lodges in the lung there is one chance in three that it initiates infection. The numbers are of course pure guesses, but the ratio between the two may be nearly correct.

The importance of these differences in the dose required to infect will be evident in other contexts but the route of infection has even greater significance for the general quality of an epidemic. Every type of infection has its own individuality and we must beware of a tendency to think always of something like measles as the norm of an infectious disease. In some ways the most instructive example to offer is wound infection. We do not normally regard bacteria like staphylococci or streptococci as being responsible for infectious disease in the conventional sense of spreading directly from person to person, yet anyone who has read the life of Florence Nightingale will know of the lethal spread of wound infection from one patient to another in the war hospitals of the Crimea. Even earlier than this Semmelweiss recognized that the appalling mortality in Viennese lying-in hospitals was the result of infection carried from the post-mortem room on the hands of obstetricians and students. The dangers of infection from the post-mortem room persisted in a different form until relatively recent years. Forty years ago if a pathologist performing a post-mortem on the body of a patient dead of some form of streptococcal infection were to prick his finger on the edge of a sharp piece of bone, he was in almost as great a danger as a man who had been bitten by a cobra. Unless the scratch was immediately dealt with by strong local antiseptics he was very liable to die as a result of infection with particularly virulent streptococci. During the fatal illness and for a short time after the death of the patient the streptococci had been multiplying rapidly in human tissues. Mutants with the ability to flourish in that environment had been selected, and when they were transferred almost immediately to other living human tissues they would usually continue to flourish, this time at the expense of the pathologist. The advent of penicillin has completely changed the situation and we no longer hear of post-mortem room sepsis.

Another type of epidemic spread which has a fundamental resemblance to wound infection is seen in relation to secondary bacterial infection in influenza, measles and other respiratory virus infections. It is one of the curiosities of military history that when large numbers of recruits were brought together, large numbers died of measles. This was very evident in the American Civil War and during 1915–16 in Australian military training camps. The measles virus as such produced more severe disease in young adults than in children, but the deaths were due to bacterial infection in lungs rendered susceptible by the virus disease. In a ward full of young soldiers with measles, streptococci and pneumococci spread easily, virulent variants were selected, and what was essentially a lethal epidemic of bacterial pneumonia developed. It is significant that when similar situations arose in 1940–1 measles deaths did not occur. The sulphonamides had arrived to deal with the bacteria. An essentially similar situation can be seen in the occurrence of the highly fatal disease, staphylococcal pneumonia. Nowadays this is the usual cause of death in those debilitated old people or infants who die during epidemics of influenza. The staphylococcus gains its foothold in a lung whose lining has been damaged by the virus. This is probably very like what happened on a larger scale in the great influenza pandemic of 1918–19. At that time the influenza virus was unknown and only the associated bacteria could be studied. Such reports as were made, however, showed that sometimes bacteria were of great importance in causing or aggravating the fatal pneumonia which was so common. The types of bacteria found in the lungs varied from one country to another, and even in different stages of the epidemic in the same country. It seems as if a very active virus swept over the whole world, finding almost all individuals susceptible to it, and in its passage made all sorts of temporary alliances with pathogenic bacteria spread by the same respiratory route. The virus initiated the illness in every case, but when a fatal outcome resulted it was almost always the bacteria which were finally responsible.

Our second major factor was concerned with the 'shedding' of organisms by patient or carrier. Perhaps the most important point to make here is that, other things being equal, the shedder will infect more people if he is still well enough to move around the community than if he is seriously ill and hospitalized. The person with influenza who persists in working and travelling because he is sure 'it is only a bad cold' is a well-known menace to the community. Equally important is the opportunity that a long incubation period can give for an infected individual to travel from the place where he was infected to a region where the disease is absent and most or all people susceptible. Transfer of smallpox or yellow fever in this fashion by passengers on jet aircraft is a well recognized modern danger, but there are many examples from former times. One of the most important effects of disease on world history was the way in which smallpox amongst the indigenes facilitated the European conquest of Mexico and later of the rest of the North American continent. The entry of smallpox is ascribed to a negro with the disease who landed with the group led by Narvaez, one of Cortez's lieutenants, at Cempoallan. The Spaniards were apparently all immune as a result of childhood infection but the disease was wholly unknown in the Americas. There is some evidence that measles, another disease new to the indigenes, played a part in the smallpox epidemic that resulted. It is impossible to know how many died of smallpox, but it was undoubtedly a significant proportion of the population, and the social and psychological effects must have been overwhelming. The long incubation period of both diseases facilitated the spread of infection throughout the country. People already unknowingly infected and seeking some distant place where they might escape from the pestilence had ten days or more in which to travel. As they became sick new foci of infection were established from which the same spread could be repeated. Similar epidemics occurred in many other parts of North and South America as European invaders spread over the two continents. Even in Australia smallpox spread disastrously amongst the aboriginals around Sydney within a few years of the founding of the settlement in 1788.

Patients hospitalized with infectious disease may of course sometimes infect those looking after them, but except in rather unusual circumstances such episodes are rare and any chain of infection normally stops at the hospital bed. Most of the exceptions concern exotic infections, contracted in most cases from animals.

In Lousiana a hunter went down with a severe fever almost certainly contracted from some bird or animal. From him there started a chain of infection involving nurses, doctors, patients and visitors that in all involved nineteen people, eight of whom died. The agent responsible was found to be closely related to that of psittacosis. A more recent story of similar happenings comes from West Africa. An American worker at a mission hospital in Nigeria came down with a severe fever and was flown to the United States for special study. The virus that was isolated from her blood, 'Lassa virus', produced fatal infection in a number of laboratory workers and all further investigations were abruptly halted. Work had gone far enough to establish that it was an arenavirus, probably derived in the first place from some rodent.

It occasionally happens that infection is derived from the body of a patient who has died in hospital. The first definitive identification of the etiology of a recent British outbreak of smallpox introduced by a Pakistani immigrant was in the pathologist who contracted the lethal disease from the body on which he was conducting an autopsy. In a London hospital a man died of an unexplained pneumonia; a fortnight later a nurse who had looked after the body and two pathologists who had carried out the post-mortem examinations came down with Q fever.

Anyone who looks at infectious disease with an eye to its evolutionary history will recognize that, in general, rapid death of the host will automatically destroy any possibility of the invading microorganism passing to another host. Apart from such curiosities as the stories we have just told, there are only one or two significant exceptions to this rule. Anthrax is one of them. In the carcass of a sheep dead of anthrax the bacillus takes on the spore form in which it can resist heat and desiccation for years. The spores from the rotting carcass may infest the ground and at any time give rise to infection of fresh sheep. Here death

123

of the host forms part of the cycle necessary for spread of the infection. A rather different example is found in regard to plague. When a plague-infected rat dies, its fleas desert the cooling body and will spread infection to any new susceptible host, rodent or man, that they can reach. In the case of plague, death of the normal host (the rat) is not necessary for the maintenance of the infection in that species, but it does play a very important part in causing human infections.

Rabies, the most regularly lethal of all virus diseases, also calls for mention here. Many epidemiologists have pondered on how rabies survives in Nature so effectively as it does – in North American foxes and skunks for example – when these hosts are regularly killed by the virus. An animal reservoir where the host is only mildly affected has been looked for and insect-eating bats considered as a possibility. Other authorities have suggested that the conventional difficulty in long-term survival of a lethal pathogen is hardly applicable to rabies. They feel that there is no need to look for a reservoir of the usual type. The unusual length of the incubation period, which may run into months, and of salivary excretion of virus before the animal becomes paralysed may be long enough to ensure its effective transfer and indefinite persistence in the ecosystem.

If we return to the general characteristics of spread of infectious disease in the community at large something must be said about the times at which infectious material is liberated from the patient or from a subclinically infected carrier. For such diseases as smallpox, measles and mumps the period of communicability begins a day or two before symptoms become severe or unmistakable. The person sickening for measles and feeling vaguely out of sorts but not ready to give up his ordinary activities is the most dangerous spreader of all. In measles the virus appears to be liberated into the secretions of mouth and nose for only a few days during the stage just before and just after the appearance of the rash, so that infection can be spread only during that period. In diphtheria the bacilli remain much longer in the throat, often for several weeks, while in tuberculosis a patient may be actively infectious for months or years. For every patient with an infectious disease we could, theoretically at least, prepare a graph to show how infectious he is for others at different periods of his illness. Most of these graphs would show a rather sudden rise from zero to a peak of maximum infectiousness usually a day or two after the beginning of actual symptoms, with a more gradual fall thereafter. The downward slope indicating the disappearance of infectiousness is on the whole much steeper for virus diseases like measles or influenza than for bacterial infections like diphtheria. The virus diseases probably make up for this by more active shedding during the brief period of their infectiousness.

Influenza and most other respiratory viral diseases are highly infectious because large numbers of virions are shed in the form of droplets which are transmitted very directly to others by sneezing, coughing or shouting. Moreover, since the incubation period of respiratory diseases is invariably short (two to four days usually) and infectious virus is liberated from superficial cells of the upper respiratory tract throughout that period and the few days that the illness lasts, epidemics of respiratory virus diseases spread through a community extremely

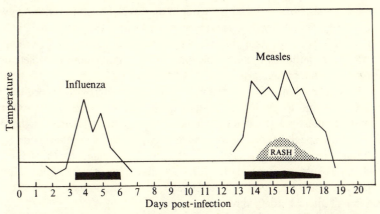

Fig. 12. To illustrate incubation period and period of communicability in influenza and measles. Symptoms and fever are indicated by the temperature curves; approximate period of virus liberation is also shown as a solid black bar. The influenza temperature is from an experimentally inoculated volunteer.

rapidly. The classical examples are influenza and respiratory syncytial virus, both of which can sweep through a major city in two months.

The spread of enteric agents on the other hand is a chancy, irregular process dependent on one or other form of faecal contamination. In general they are readily controlled by personal cleanliness and good plumbing. When standard precautions break down, however, enteric organisms have the opportunity of infecting large numbers of people simultaneously, in so-called 'explosive' epidemics due to faecal contamination of communal supplies of water, milk or food. Furthermore, enteric organisms tend to be excreted over much longer periods of time than those of the respiratory tract. The most notorious of all carriers of infection are the people who on recovery from a bout of typhoid fever develop a persisting but wholly symptomless colonization of their gall bladder by the typhoid bacillus and are a potential danger to their associates for many years. For some reason typhoid carriers tend to be middle-aged women and if they are cooks or food handlers they are naturally more prone to infect others and be recognized as carriers. The most famous (or notorious) carrier was an American cook, 'Typhoid Mary', who in ten years cooked for eight different families and was apparently responsible for seven epidemics involving more than 200 persons.

Before concluding discussion of these multiform aspects of the process by which infectious material is shed into the environment to reach a new susceptible, some mention should be made of the varying capacity of microorganisms to retain their infectivity outside the body. How long can a given pathogen remain alive and infective in air-borne droplets or faecal material, particularly when these dry out?

Unlike the highly fragile respiratory viruses, enteroviruses are relatively resistant to drying and other environmental hazards. Another group of intestinal organisms, bacilli of the genus *Clostridium,* have evolved the special capacity to survive long periods of drying in soil by forming highly resistant spores. There are documented cases of footballers or soldiers contracting fatal tetanus from *Clostridium tetani* spores excreted years earlier in horse manure that found its way on to the playing arena or battlefield respectively.

Droplet infection is not by any means always a simple matter of producing droplets by coughing or sneezing and the immediate inhalation of such droplets by the next susceptible individual. In both World Wars crowded camps and barracks showed high incidences of some of the typical 'droplet' diseases, especially meningococcal meningitis and the streptococcal throat infections which sometimes lead to rheumatic fever. In the First World War it was shown that the prevalence of meningitis was directly related to the degree of crowding in army huts used as sleeping quarters, and it was thought that the reason was simply that the infectious droplets had to pass a shorter distance from carrier to new susceptible host. In the Second World War the most popular interpretation of the spread of infectious disease in sleeping quarters was quite different. The floating droplet was regarded as of minor importance in comparison with contaminated bedclothes and floor dust. By one means or another saliva and nasal secretion can easily contaminate blankets or other textiles, and when these are shaken, clouds of bacteria are liberated into the air and eventually settle as dust. The inhalation of dust raised during bed-making and sweeping was the important factor, and special techniques for oiling floors and treating bedclothes rapidly became standard hospital practice.

*

Most of the factors concerned with the final stages of transfer of symptomatic infection to new hosts have either been dealt with in Chapter 8 or are so little understood and variable as to be unmanageable. Our third and fourth factors of probability of contact and extent of pre-existent immunity are best illustrated by a quantitative analysis of the way in which epidemics develop.

With so many complex and variable conditions to be considered, it is no wonder that the only practicable method of treating the available data about the prevalence of disease is statistical. Detailed investigation of particular instances of infection may often provide information of much value, but chance plays so great a part in determining whether, when, and how severely any individual is attacked that any general conclusions must be based on large numbers of experiences, treated according to proper statistical methods. In most civilized countries statistics are available as to the numbers of deaths from all the infectious diseases, and for the numbers of cases notified of certain scheduled diseases, usually the more serious infections. As a rule, the age and sex distributions of cases or deaths are also available. For particular outbreaks that have aroused public interest, it is generally possible to obtain weekly totals of cases or deaths, and to ascertain in what municipalities or other administrative districts they occurred. In general, this information is the raw material from which the epidemiologist must draw his generalizations and hypotheses.

An epidemic in the popular sense of the word is merely a prevalence of a particular type of infection which appears to be unusually concentrated in time and space. Childhood diseases like diphtheria and measles in the days before immunization came in epidemics or prevalences with intervening periods in which the disease was either absent or at least significantly rarer. When the data are arranged in the form of a graph with time along the horizontal axis and the number of cases or deaths shown vertically, we have for nearly every endemic disease a more or less regular wave form, the crests of the waves representing the times of epidemic. Fig. 13 shows examples of such wave-like curves for measles and diphtheria in an Australian community.

In these oscillations we have a reminder that although we can legitimately talk of a long-term balance between host and parasite, such equilibria were far from evident in short-term day to day experience even when there was no deliberate and effective human action to interfere with the process. It is possible that in the early years of this century infection with tuberculosis, which is a very slowly developing insidious disease, may have reached a stable state in which it spread at just about the rate required to infect each month almost all the susceptible individuals entering the community by birth or immigration. Yet there is much to suggest that even the tubercle bacillus moved irregularly, and localized epidemics of primary infection took place especially in schools. When a more typical epidemic is at its height there are large numbers of individuals who can serve as sources of infection, and there is a high

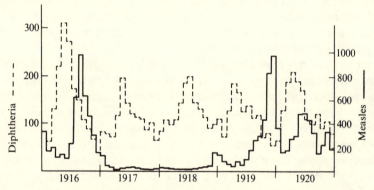

Fig. 13. Waves of prevalence in two endemic diseases, measles and diphtheria. The monthly notifications of measles (heavy continuous line) and diphtheria (broken line) in South Australia over a period of five years.

probability that any susceptible will contract the infection. With exhaustion of susceptibles, the epidemic dies down rapidly, and what cases occur will have little chance of passing on their infection. The rate of spread of an epidemic at any moment is not only a function of the number or density of susceptible persons available, but of the number of sources of infection as well. An epidemic will affect some of even a small proportion of susceptibles if the number of infecting individuals is large, but if there are very few sources of infection, the chances of the same small number of susceptibles escaping infection are more than proportionately increased. Perhaps a diagram will make this clearer. In Fig. 14 are shown the lines of infection from one case to another at various stages of a hypothetical epidemic, starting in a wholly susceptible population. The incubation period is taken as a week, and three four-weekly periods are shown. It is assumed that once an individual has been infected he becomes immune, and underneath are shown the proportions of susceptibles and immunes at each period. In the first month, one case at the beginning gives rise to ten at the end of the period. At the peak of the epidemic many cases have no opportunity of infecting others, but there are still the same number of cases at the end of the period. As we approach the end of the epidemic, very few lines of infection can be continued, for all the contacts of most cases are already immune. If the epidemic is one like measles or influenza in which the period over which an individual is infectious is limited to a few days, the

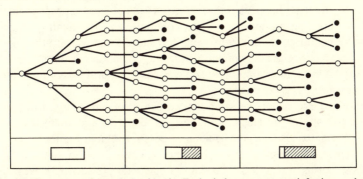

Fig. 14. The course of a typical epidemic. Each circle represents an infection, and the connecting lines indicate transfer from one case to the next. Black circles represent infected individuals who fail to infect others. Three periods are shown, the first when practically the whole population is susceptible, the second at the height of the epidemic, and the third at the close, when most individuals are immune. The proportions of susceptible (white) and immune (hatched) individuals are indicated in the rectangles beneath the main diagram.

disease will die out completely in a community when the proportion of susceptibles reaches a certain low level.

Occasionally an opportunity arises by which this process can be followed in detail. This happened in Melbourne when a new type of influenza appeared in 1957. Each week over the whole duration of the epidemic Keogh arranged for random samples of blood to be put aside from 100–200 blood donors. Subsequently the serum was tested for antibody against the new virus. There were no positive results in the first week, but thereafter the percentage rose regularly in the expected fashion but flattened out when approximately 45 per cent of donors had acquired the new antibody. The epidemic had apparently died out when about half the susceptible adult population was still uninfected. Of course adult blood donors cannot be considered a representative cross-section of the community. Less comprehensive studies of the incidence in some groups of school-children during the same Melbourne outbreak showed that a much higher proportion were infected, indicating that school and family contacts were clearly more effective than casual adult encounters. Nevertheless the message is clear. The popular view that a widespread epidemic of influenza affects 'virtually everybody' is quite fallacious; the epidemic burns itself out before that has happened.

It is quite possible for a large city like Melbourne to be completely free from epidemic influenza or measles for two or three years. Of course, epidemics of such diseases do not all occur simultaneously in different places, and what happens is that there is a kind of irregular peregrination of the infection around the various communities of a country, or, now that all countries are linked by rapid transit, of the world. By the time it returns to our original community, a fresh population of susceptibles has arisen by birth, migration or, in some diseases, natural loss of immunity, and the process is repeated.

When there is not the same limited period of infectivity, a different state of affairs develops, and carriers play an important role. In diphtheria, for instance, the bacilli may persist for months after infection, particularly if the nasal cavity becomes infected. If a few such carriers persist during the period following an epidemic prevalence, even if the proportion of susceptibles is low, a few cases will keep on occurring; there will therefore be no such striking ups and downs as there are with measles and influenza, and no periods in which a large community is free from infection. The carrier state is almost restricted to bacterial and protozoal infections. It is rarely seen with viruses.

There is much more that could be said about the wave-like shapes of any graph of infectious disease prevalence such as Fig. 13. Seasonal factors undoubtedly play some part in determining, other things being equal, when a new epidemic wave will appear. But waves of any sort need more than just the interval between their crests to describe them fully. We must also know the height of the waves above the troughs and something about the shape of the wave. This applies just as much to the conventional waves which appear in a graph of the prevalence of an endemic disease as to the waves of the sea. We have discussed why increased prevalences occur periodically, and what factors influence the times at which they occur. We have yet to deal with what determines how big the epidemic will be when it does occur.

One of the important factors will obviously be the number of susceptible individuals present at the beginning of the prevalence. If instead of coming every second year measles remains absent for four or six years, then when it does recur the number of susceptible children available, and hence the number of cases, will be proportionately greater. During the earlier history of Melbourne, measles recurred at rather long and irregular intervals, and the number of deaths in each epidemic showed a rough proportionality to the length of the interval during which the virus had been inactive. Only statistics of deaths were available in those days but they can be accepted as roughly proportional to the number of cases or to the number of children non-immune at the beginning of the epidemic. Chickenpox, mumps and rubella show very much the

Table 3. *Measles epidemics in Melbourne*

Epidemic	1866–7	1874–5	1880	1884	1887	1893	1898	1900
Previous free interval (years)	6	7	4	3	2	5	4	1
Number of deaths	433	733	174	176	64	386	403	95

same epidemiological behaviour as measles, with the proportion of non-immunes in the community playing a dominant role. The special capacity of rubella to damage the embryo in early pregnancy has been discussed in the previous chapter where something was also said about the 'second face' of chickenpox, herpes zoster.

This discussion of the common infectious diseases can be summarized by saying that the nature of epidemic prevalence and the age incidence of attacks depend on (*a*) the quality of the infective agent, (*b*) the route by which it is spread and (*c*) the concentration of susceptible individuals, i.e. of those not already immunized by experience of infection or by an appropriate vaccine. Broadly speaking, once a child has been infected he remains almost completely immune so long as he remains in the community. It does not necessarily follow that the first infection alone is sufficient to confer lifelong immunity. There is the important alternative that the first infection supplies a basic immunity which requires to be periodically reinforced by minor reinfections if it is to remain effective.

The duration of immunity following a single attack undoubtedly varies very greatly from one type of infection to another. The most clear-cut evidence on the matter is obtained from the experience of isolated island communities. Much of the interest in island epidemiology depends on the fact that in small, isolated communities very few types of infection can maintain themselves indefinitely. The community may therefore remain untouched by some disease for very long periods and the pattern of infection when it reenters will often provide revealing contrasts which highlight the significance of immunity.

The Faroe Islands in the North Atlantic have been inhabited for some centuries by a civilized Danish community. It was always clear to their physicians that the behaviour of infectious disease in these islands was very different from the usual European experience. The epidemiological happenings tended to be much more dramatic, and it was natural that they should be carefully described. These records, extending over a century and a half, have provided much valuable information for epidemiologists. They showed unmistakably, for instance, that the immunity conferred by an attack of measles is of lifelong duration. After a widespread epidemic in 1781, there was a complete absence of measles from the islands for sixty-five years. When it returned in 1846, the whole population, except for a few greybeards who had been infected in infancy, went down with the infection. The next epidemic appeared in 1875, and on this occasion accurate records showed that only persons who had not been affected by the previous epidemic, i.e. persons under thirty, were susceptible, and that approximately 99 per cent of these had typical measles.

We can hardly call measles a typical infectious disease. It is almost the only one in which, with insignificant exceptions, every susceptible person, irrespective of age, who becomes infected shows the typical symptoms of the disease. The permanent immunity which follows infection with measles is not unique, but it is much commoner in other diseases for immunity to be temporary and to disappear almost entirely after a variable period if it is not occasionally reinforced. Isolated communities offer many examples of this state of affairs.

It can be stated almost categorically that any community which is cut off from the rest of the world for periods of a year or more will, on the arrival of visitors, suffer an epidemic of illness of the feverish cold-influenza type. A detailed study of this phenomenon was made at Spitzbergen in the years just before the Second World War. On the island there was a moderately large resident population engaged in coal mining who, for approximately nine months of the year, were completely isolated from the outside world. During the height of the winter the three communities were also isolated from each other for approximately three months. Soon after the arrival of the first ship in summer an epidemic of respiratory infection, always colds, sometimes 'influenza', would appear and move through most of the population before fading away in the late autumn. During the winter night the population remained free from colds and other infections. With the spring the mutual isolation of communities on the island was broken down and with the mingling of the groups minor outbreaks of mild colds sometimes appeared, but no serious prevalence until the next ship from the outside in July or August.

Such occurrences were naturally commoner in days when rapid transport was less universal. The 'stranger's cold' which affected the islanders of St Kilda off the west coast of Scotland whenever a ship touched the island was commented on by Dr Johnson, and was observed by medical writers several times in the nineteenth century. Similar phenomena are on record about the Faroes, Tristan da Cunha and others of the proverbially lonely spots on the globe. Antarctic expeditions have suffered similarly with the arrival of a relief ship or on their return to civilization.

These outbreaks of mild infections occur even when the people who bring them appear to be perfectly healthy, and it is characteristic that the strangers are themselves unaffected. Any interpretation will be based on the certainty that in any large community there is a constant interchange of the viruses and bacteria which can occupy the upper respiratory tract. There are at least 100 different viruses that can cause colds, and by different we mean antigenically different. A serum from someone who has recently recovered from cold No. 1 will neutralize rhinovirus type 1 but not types 2 to 100. It is not humanly possible to keep track of all the viruses but limited studies of colds in student groups indicate that after having a type 1 cold the subject's colds over the next couple of years will not include any due to rhinovirus 1. This means that there is some degree of type-specific immunity that lasts at least two years. In the past the commonest interpretation of the epidemiology of the common cold was that for each of us in a city there was a continuous invisible process of inapparent infections with an ever-changing pattern of partial immunities based on the liberation of IgA antibodies on the nasal mucous membranes. Occasional breakdowns or gaps in immunity allowed our average two and a half colds per annum.

With the discovery of interferon an alternative explanation has been proposed. The reason we go for some months between colds, on this view, is because the interferon produced in the nose by one type of rhinovirus protects against infection by any other virus for several weeks. This is not specific immunity but a sort of short-term 'broad spectrum' resistance which could be important in protecting us from an endless sequence of respiratory infections. It provides also a neat explanation of the susceptibility of the residents of Spitzbergen after a long winter of isolation from the outside world. Throughout the winter no rhinoviruses have been circulating amongst the local residents, who therefore have neither IgA nor interferon to protect them when the first ship arrives in spring. Whatever rhinovirus happens to be introduced by the first spring visitors will spread rapidly within this tightly knit community with a near 100 per cent morbidity rate. The resulting IgA will protect the people against that particular rhinovirus for the next few years but it is virtually certain that the next spring will see the introduction of a different serotype. Opinions differ about many aspects of the common cold; some think that allergic factors also play a part and the significance of superadded bacterial infection has been incompletely investigated. The situation may turn out to be too complex to analyse completely. Almost certainly both interferon and immunity play their parts and it is probably not very important to define their respective roles. It will be of much interest to watch the results on the incidence of common colds, of the attempts to induce interferon production in volunteer subjects that will undoubtedly be made in the next year or two. They may establish the importance of interferon – or swing our thoughts back to IgA antibody.

In the history of epidemics there is a particularly interesting series of outbreaks which created great alarm and excited much interest in England during the Tudor period – the English Sweats. Although the nature of the microorganism causing the disease is unknown, we can feel reasonably confident that it was a virus, and probably one spread by droplet infection. The behaviour of the disease exemplifies a number of interesting epidemiological points, and it is worth while giving a fairly full account of the five outbreaks. The information is taken almost entirely from Creighton's *History of Epidemics in Britain,* which is the classical and almost the only account of the history of disease in Great Britain.

The disease appeared suddenly in London in 1485, a fortnight after the arrival of Henry VII fresh from his victory over Richard III at Bosworth Field. The symptoms were highly characteristic, a sudden onset with 'great swetyng and stynkyng with redness of the face and all the body and a continual thirst with a great heat and headache.' Death sometimes occurred within one or two days from the onset, but many, probably the great majority of those infected, recovered after a brief illness. The infection attracted particular attention by its tendency to affect members of the upper classes of society rather than poorer folk.

On its first outbreak in London, two successive Lord Mayors died of the Sweats within a month, as well as several aldermen. In 1517, during the third epidemic, Cardinal Wolsey suffered a severe attack with relapses, but recovered, while in the course of the next outbreak (1528) Anne Boleyn was infected, but apparently had only a mild attack. Although little or no mention is made in contemporary writings of the incidence of the disease on the lower classes or on children, there are distinct implications that both groups were little affected in comparison with adults of the wealthier classes. During the epidemic of 1528 the disease spread to the Continent, and showed a curious tendency to avoid the French and spread freely amongst the German kingdoms. It is said that when the Sweats broke out in Calais, then an English possession, only the English were affected, the French were untouched. Even if the effect of these differences of race, class and age on the incidence of the disease had been exaggerated, there can be little doubt that they were in the direction indicated.

In his discussion of the nature of the English Sweats, Creighton put forward a theory which, with slight modifications, appears to provide an adequate account of the happenings. Henry Tudor invaded England in 1485 with troops, nearly all of whom were French mercenaries drawn from the valley of the Seine. From this region there are records of occasional outbreaks, particularly in the eighteenth century, of a relatively mild infectious disease with symptoms resembling those of the Sweats. Some of the French soldiers were apparently carriers of the virus responsible in France for these mild endemic infections. Most of the others must have been immune.

In England the virus appears to have been present, but in a less virulent form, and spreading easily only under crowded and filthy conditions. When Henry VII's troops reached London, carrying what was for England a particularly virulent strain of virus, two more or less distinct populations were exposed to its attack. On the one hand were the poorer classes, amongst whom there was a high incidence of all infectious disease and an enormous infantile mortality. On this section of the community the Sweats made little impression. No doubt there were fairly numerous cases, but there was a sufficient degree of immunity to prevent the disease being severely felt. The immunity may have resulted from past infection with a relatively avirulent strain of the same virus, or possibly with other viruses having some antigenic similarity to that of the Sweats.

Infectious disease was prevalent enough amongst the upper classes, but there was not the same constant exposure to all sorts of infections as amongst the less fortunate. No immunity was apparently present against the virus, and for this part of the population the Sweats represented a new disease. As far as can be ascertained, the incidence of the Sweats fell chiefly on adults, children being very little affected. Probably adults of all ages were more or less equally liable to the disease, but the chief mortality was amongst those past middle life.

The apparent insusceptibility of the French to infection is obviously to be explained by the supposition that the virus was widely endemic through the country, but had not previously found conditions suitable for spread outside of France. Whether or not the so-called 'Picardy sweats' of the eighteenth and early nineteenth centuries were the lineal descendants of this endemic French virus will probably never be known. At the present time there are no known infections which can be regarded as of the same type.

If this interpretation of the English Sweats is correct, we have a striking example of how in one country a disease may be endemic and produce a widespread immunity with a minimum of visible disease, while in another country, where such immunization has not occurred, its introduction results in outbreaks of serious and often fatal disease. On a more dramatic scale, it is merely the story of the 'stranger's cold' again.

Of course there is also the converse of the picture by which the stranger coming from a non-endemic region falls an easy victim to diseases which seem to have no effect whatever on the natives. The story of yellow fever in the West Indies or on the West Coast of Africa is the classical example, but that is best left for a chapter of its own. The history of military operations by European powers in the tropics contains many other instances of the same general type right down to the 1939–45 war. In New Guinea malaria and scrub typhus were the two important medical diseases amongst Australian and American army personnel. Neither caused any evident effect amongst the adult native population provided they had been brought up in the endemic areas. Malaria, however, was just as serious for Melanesians from the non-malarious highlands brought down to the coastal regions as for the white Australians in the army.

10. Evolution and survival of host and parasite

It is hard to escape from the anthropocentric attitude that epidemics and prevalences of disease are purely human problems to be controlled and eliminated by human action. In a sense that has been the justification for the professional careers in microbiology that we and our colleagues have chosen. But it has never been the central theme of this book since it was first written over thirty years ago. That theme has been and remains the evolution of the host–parasite relationship and the understanding of current ecological situations in evolutionary terms. At this point in the discussion of things as they are from the public health man's point of view, it is expedient that we return to a more biological consideration.

Evolution is a process of continuing change that may have started $3\frac{1}{2}$ thousand million years ago and will continue 'till the rocks melt in the sun' and the earth is finally sterilized aeons ahead in the future. In terms of individual lives even of long-lived mammals like ourselves the rate of change is excessively slow. Until about a century ago it was only very rarely that a significant ecological change would take place during a man's lifetime. The only important exceptions were when virgin land was being opened up for agriculture or when an old established culture was destroyed by violence as in Mexico and Peru. In the days before agriculture and organized war, when men were nomadic food gatherers, we were as much part of the ecosystem as any other species. One might say that until modern man came to dominate the planet, the standard situation everywhere was an interlinked set of balanced ecosystems each with a character determined primarily by the physical quality of the environment and within which a range of living species – plant, animal and microorganism – managed to survive and to maintain roughly constant numbers. In a sense each balanced ecosystem was a way-station on the path of evolutionary change.

In a favourable environment bacteria and viruses can replicate themselves in minutes or hours. Huge new populations can be built up from a single survivor in a matter of days, and the potentiality of changing inheritable qualities rapidly when the environment demands it is the great evolutionary asset of bacteria. If the relevant environment remains constant, however, the general rule of an approximately stable equilibrium holds even for microorganisms parasitic on higher forms. They are as much a part of the balanced ecosystem as the green plants, the herbivorous mammals and the predatory carnivores.

In this chapter we are interested primarily in the parts played by potential agents of human or animal disease in such balanced ecosystems. In most cases it will be found that our interest has been provoked because of what happened when modern man intruded into an ecosystem within which he had not evolved. In Africa however, where man himself has been evolving continuously since the days of Australopithecus two million years ago, there are illuminating differences. Man as a wandering hunter and food gatherer can be integrated into a natural ecosystem as readily as any other mammalian species. Only when he becomes agriculturalist, city dweller and industrialist does he play havoc with the ecosystems. That havoc has repercussions of various sorts on man and his infectious diseases and in the final chapter these will be discussed. Even at the present ecological level presentiments of the biological absurdities of human behaviour will come seeping in.

To the observing biologist the web of life – of which he himself is part – has no obvious point in time, space or species at which to make a logical beginning to its description. Every organism is itself a product of evolution and ecologically related to almost every other organism in the ecosystem. As an arbitrary entry into our theme we have chosen to consider the first impact of an infection on a population, human or animal, which has had no previous experience of that disease. In the history of human infectious disease there are some famous examples; the conquest of Mexico by smallpox (rather than by Cortez), and the fatal impact of measles and tuberculosis on the islands of the Pacific in the nineteenth century may be instanced. None of these historic examples however could be studied according to modern standards of epidemiological research and deductions from contemporary medical and other records are necessarily highly speculative. We shall return to the human problems of such 'virgin soil' epidemics after we have discussed the outstanding modern example of the introduction of a wholly new infectious disease to a very large population of a wild species. It was an episode which from the ecologist's angle had the enormous advantage that it could be conveniently studied both in the field and in the laboratory – myxomatosis of rabbits in Australia. Our discussion of this landmark in the history of epidemiology will be wholly based on Fenner and Ratcliffe's book *Myxomatosis*.

The virus was first isolated in Montevideo from an outbreak of fatal disease in laboratory rabbits of the common European species *Oryctolagus cuniculi*. Its real origin remained a mystery until 1942 when Aragão recognized that the wild rabbits (*Sylvilagus*) in the São Paulo region of Brazil were naturally infected and formed a reservoir of infection from which the virus was transmitted to other rabbits, wild or European, by mosquitoes. The association of the Brazilian rabbit with myxomatosis has all the marks of an ancient and stabilized equilibrium. Instead of a fatal infection, the bite of an infected mosquito provokes only a small swelling of the skin and a transient invasion of the blood by the virus. The rabbit survives and is subsequently immune even to the minor results of primary infection. The European rabbit is only a distant relative of the Brazilian, and for it, infection with myxomatosis virus is the same kind of biological accident as scrub typhus is for a human being. The disease is almost always fatal.

Fig. 15. Wild rabbit showing the typical facial swelling and half closed eyes of myxomatosis. (UK Ministry of Agriculture and Fisheries.)

Australia's traditional animal pest is the European rabbit, which in 1859 was effectively introduced into an area near Geelong by 'an ardent acclimatizer', Thomas Austin. The animals flourished and at a rate of around seventy miles a year spread over all the areas of the continent that were capable of supporting them. From the beginning the rabbit was regarded first as a potential danger, then as an established pest. As early as 1888 Pasteur had suggested that a bacterium (*Pasteurella*) should be tested as a means of control and the use of myxomatosis virus was considered from 1919 onwards. It had two quite unusual and desirable characteristics for such a use – its extremely high mortality rate and its strict limitation to rabbits. It is innocuous even when deliberately injected into human beings, domestic mammals or native Australian marsupials. Early attempts to initiate infection in wild rabbits were essentially unsuccessful. All the rabbits in a warren might die, but the disease spread no further. In these experiments the aim was to allow the disease to spread by contact from rabbit to rabbit, and it so happened that the tests were carried out in areas virtually free from mosquitoes. In 1950 experiments were begun in a different part of Australia near the upper Murray River and again the results looked unpromising. Then just before Christmas a wholly unexpected development occurred. Rabbits were dying, obviously from myxomatosis, on the river flats near Albury fifteen or twenty miles from the nearest rabbit warrens that had been experimentally infected. In the next three months myxomatosis spread rapidly over an area about as large as western Europe, concentrating its lethal action along waterways but sweeping north across New South Wales in leaps that seemed to be as much as a hundred miles at a time. The mortality amongst rabbits within the vast area involved was very patchy, but along the river flats there would often be virtual extermination and in areas of western New South Wales, where unusual rainfall had left much water lying, almost equally good results were obtained.

The virus could be spread by any mosquito or other biting insect that will feed on rabbits, the dominant carrier varying from one season to another. *Aedes* and *Culex* were important in the first summer, but in 1951–2 anopheline mosquitoes played the major part in another even more extensive outbreak over most of southern Australia.

In subsequent years myxomatosis persisted over the whole area, rising to a peak of activity whenever a summer prevalence of mosquitoes coincided with a recovering population of rabbits. From the first it was obvious that the disease provided a special opportunity for the epidemiologist. It was under study from its inception and provided a unique opportunity to follow the process by which a virus and its host can reach an equilibrium.

By 1957 both the virus and the rabbit population had changed. Within two or three years after its initial establishment the myxomatosis virus being isolated in the field was of lower virulence. Instead of 99 per cent of deaths the figure was nearer 90 per cent and, more important, the time between infection and death was almost doubled. This new form of virus was shown to survive much more effectively in the wild than the original extremely virulent form, or than the rare forms of much lower virulence

which are sometimes isolated. Change in the resistance of the rabbit
population was slower in developing, but by 1957 it was clearly evident.
By this time there were areas in which rabbits had been exposed to five
successive epidemics, each at least 90 per cent lethal. This is a very
stringent test for selective survival, and in the last set of tests for
resistance amongst young rabbits from these areas the results were very
striking. Using a virus killing 90 per cent of normal rabbits, the mortality

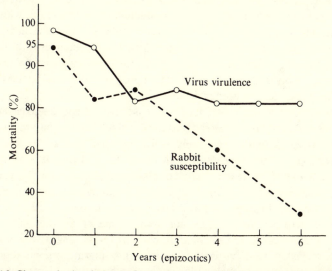

Fig. 16. Changes in the virulence of myxoma virus and in the genetic susceptibility of
rabbits in Australia since 1950. Both are measured in terms of per cent mortality. Virus
virulence represents the average mortality conferred by strains recovered in the field after
the number of years shown on the x-axis, and tested on standard laboratory-bred rabbits.
Rabbit susceptibility is that of young rabbits caught in areas which had experienced the
number of epizootics shown and challenged (after any maternal immunity had vanished)
with a virus strain of moderately high virulence (one isolated about a year after first
liberation of the virus and found to kill 90–95 per cent of normal laboratory rabbits).

in the group that had experienced all these epidemics was just less than
50 per cent. Many of the surviving rabbits suffered an illness that in the
wild would probably have killed them. It should be stressed that the
recovery of these rabbits was not due to any specific immunity acquired
as a result of previous exposure to myxomatosis. They had in fact never
encountered the virus, though their ancestors had. Their changed resistance was innate and inheritable – an example of natural selection in a
very intensive form acting to favour gene mutations that in one way or

another conferred a greater chance of surviving infection. Even so it was by this time obvious both to the investigators and to Australian farmers that 'myxo' was not going to eradicate the rabbit.

In Fenner's appraisal of the situation in 1965 he stated that myxomatosis had persisted in virtually all areas and at intervals still produced 'good kills' amongst the rabbit population. Rabbits were relatively resistant but although infected animals would survive in a laboratory environment most of them would be easy prey for predators – foxes, feral cats and eagles – in the field. At a guess the rabbit population and myxomatosis virus have reached a fluctuating equilibrium with the rabbits at around 20 per cent of their numbers before the advent of myxomatosis. The whole story underlines the principles (*a*) that a pathogenic agent may be too virulent for its own optimal survival and (*b*) that when exposed to an intense selection pressure an animal species may develop a genetically based resistance.

The history of the transfer of a disease from a species (*Sylvilagus*) with which the virus had already attained a stable equilibrium to another species (*Oryctolagus*) is probably still unfinished. It may be that in a thousand or a million years, if both the rabbit and the virus persist in Australia they may reach a greater mutual tolerance than the present fluctuating balance. Nothing analogous to the impact of myxomatosis is known in the history of infectious disease in man. One wonders, however, whether Africa may not at one time or another have seen something similar in the interaction, perhaps for as long as two million years, of yellow fever virus and the evolving, migrating populations of early hominids and their fully human descendants. Unfortunately, although a great deal is on record in regard to the distribution of yellow fever antibodies in Africa, there are only indirect hints of a genetically based resistance in the people of the endemic areas. In general the relative shortness of the times – in terms of generations of man, i.e. twenty-five year intervals – since the various human groups became geographically isolated from one another ensures that there are no great genetic differences in resistance between races. The response of virgin populations of man is attributable to their absence of specific immunity acquired in childhood rather than to any genetic susceptibility. Indeed it is by no means always true that a virgin population necessarily suffers a calamitous epidemic on its first encounter with a potentially dangerous virus or bacterium. One outbreak of polio type 3 in the Aleutian Islands showed a number of persons with symptoms but no cases of paralysis, although the investigators could show that none of the population had had any previous exposure to this virus. Even in the classical instance of a virgin soil outbreak, the measles epidemic of 1875 in the Fiji Islands in which between 20,000 and 40,000 people died, the deaths were due largely to the complete social disorganization and shattering psychological effect of the epidemic rather than to any special racial susceptibility to the disease, and since 1875 epidemics of measles in the Fiji Islands have been essentially the same as in any other parts of the world.

*

It is more relevant to our general theme to consider situations where man intrudes into a more or less balanced ecosystem involving wild animals and potentially pathogenic microorganisms. There are two groups of infectious disease that show various stages in the process by which disease in one species can be transferred to another species, the second species in both cases being man. These are first, the rickettsial diseases, of which typhus fever is the best known; and secondly, the trypanosome infections spread by tsetse flies in central Africa.

As we outlined in Chapter 5 the rickettsiae can be regarded as tiny bacteria which have become so specialized for living within the cells of their hosts that they cannot be made to grow apart from these susceptible cells. Almost certainly their evolution as parasites started in insects and ticks, many species of which still contain rather similar parasitic microorganisms, which are not, however, pathogenic to vertebrates. It was probably at first a sheer accident that one of these tick parasites should prove to be capable of living and multiplying in the rodent on which the tick fed. But once the transfer had been accomplished, a *modus vivendi* between the rickettsia, the tick and the rodent had to become established. A typical example of such mutual adaptation has been revealed in the course of investigations on the spotted fever of the Rocky Mountain States in America.

In the state of Montana there is a valley of evil reputation, the Bitter Root Valley, where since the country was first opened up in 1880, prospectors, surveyors and settlers have been liable to a severe fever, with a mortality of about 80 per cent. Except for pneumonic plague and rabies it is probably the most fatal infection known. It was soon discovered that the disease followed tick bite and was due to a rickettsial organism present in the tick. The natural history of the disease and of the ticks which spread it was elucidated in the Montana laboratory of the United States Public Health Service, and eventually a method of preventing the disease was worked out and shown to be practicable. The danger of this work was extreme. Before the method of protective immunization had been developed, seven workers in the laboratory had contracted the

143

disease. In every instance it was fatal. Since the immunization of all workers has been a routine, fifteen more cases have occurred in the laboratory, with one death.

Ticks have rather a complicated life history with three main stages, larva, nymph and adult. The tiny larvae are hatched on the ground, and when opportunity offers, attach themselves to some small rodent. For a day or two they slowly engorge themselves with blood and then drop off on to the grass. Here they digest their meal, grow and prepare for the moult to the nymph stage. The nymph repeats the same process, also feeding on small rodents, but when its engorgement is complete it hibernates through the winter and emerges as an adult with the melting of the snows the following spring. The adult needs a larger animal to feed on, bears and wild goats being the chief native hosts, but cattle, horses and man are equally suitable. The normal life history of the rickettsia of spotted fever probably involves an alternating sequence of infection in ticks and one or other of the small rodents of the region. The larval tick feeds on an infected rodent and takes in the rickettsia with its meal of blood. The rickettsiae multiply in the tick tissue and establish a harmless infection which persists through the life of the tick. With each subsequent feeding the host is liable to be infected by the bite. In the nymph stage the disease may be passed on to a new susceptible rodent; in the adult stage the normal hosts of the tick are insusceptible to the rickettsia, and it is only by an unfortunate accident that man happens to be so susceptible. The important business of passing on the infection from one natural host to the other is done in the larval and nymphal stages of the tick. The fact that the adult tick bite is also infective for man is a mere accident of no significance to the survival of the rickettsia.

The natural disease among wild rodents has never been observed. The only evidence that it exists is that a proportion of the wild rodents show immunity to the disease, of the type one would expect to follow natural infection. It is certain that, like the disease resulting from artificial inoculation of the rickettsia in the same species, the natural infection is an extremely mild one. Long experience has made it certain that this rickettsia is never responsible for the outbreaks of disease which occur when rodent populations increase abnormally. It is just possible that the rodents play no essential part in the disease, since a proportion of larval ticks hatched from the eggs of infected ticks are also infected. If this occurred regularly we might regard the whole condition as a tick disease only, the infections of rodents and man being both unnecessary accidents. However, most workers are convinced that the tick-rodent–tick cycle is much more important than the occasional transmission of the infection from infected female ticks to their offspring.

So far we have been speaking of the extremely virulent spotted fever of the Bitter Root Valley. A curious feature of the disease is that in most other parts of the United States the mortality is very much smaller and may be as low as 5 per cent. Amongst the sheepmen of southern Idaho, not very far from Bitter Root Valley, it is common enough, but is regarded as hardly worth worrying about. The rickettsia is antigenically the same in both parts, as far as can be judged by cross-immunity tests; the only difference between the strains is in their virulence. In the Eastern states where the disease has now been recognized with the domestic dog as the chief reservoir of infection the mortality in man is around 25 per cent. In all probability these differences in virulence for man are a wholly accidental property of the rickettsia which is quite irrelevant to its continued survival in nature.

On the opposite side of the Pacific there is another rickettsial disease which is present from Japan southward through Taiwan, the Philippines and New Guinea to North Queensland. Further west it is common in the Maldive Islands and Ceylon, patchy in India, and from Burma it is found continuously through Malaya and Indonesia to the Solomon Islands. Its Japanese name was *tsutsugamushi* fever, but elsewhere it is usually called scrub typhus. It is a severe disease with a mortality that varies like that of Rocky Mountain spotted fever from one infected region to another and may be as high as 50 per cent or as low as 1 per cent. Like the American disease, scrub typhus is contracted away from human communities, not in mountain valleys but on tropical islands, in grassy clearings at the jungle edge or in any other situation where the mite that carries the infection can flourish. During the war in the south-west Pacific, scrub typhus ranked next to malaria as the most important cause of disability and death from infectious diease. Nearly 7000 cases with 284 deaths were reported in the American forces. The carrier is one of two very similar mites which, in their larval stage, may obtain the single blood meal needed for their development from some small animal or from man. The rickettsia is never passed directly from man to man or from animal to animal. After its blood feed the larva becomes successively nymph and adult which live in the soil, and when eggs are laid by an infected female they contain rickettsiae which, in turn, develop in the new generation of larvae. The human infections play no part in allowing the survival of the rickettsia and its human virulence must again be regarded as a mere accident.

Typhus fever in the classical form, which has been the great attendant of war and famine in Europe for centuries, is a purely human disease. It is a severe fever with a characteristic rash, highly fatal in susceptible adults, particularly under famine conditions, but as a rule affecting children less severely. The body louse is the vector from patient to patient. It takes up rickettsiae from the blood and is itself fatally infected in the process. However, it has about a week in which to transfer the infection to another subject before it dies. In view of this method of transmission, typhus can flourish only in circumstances of poverty, overcrowding and filth. In the English history of disease, typhus figures mainly in connexion with the Civil War and as a disease of prisons, though it was not uncommon in the slum quarters of cities until the middle of the nineteenth century.

There is one feature of typhus which suggests that it is not an old-established human disease, its fatality for the louse. This is one of the few instances in which the insect vector of a human disease is seriously discommoded by the organism it transmits. The logical inference is that typhus of the louse is a relatively recent development, and that the disease has evolved in other hosts than man and the louse. The first indication that there were other methods of contracting typhus than by louse bite came from observations made in Australia soon after the First World War. Hone in Adelaide saw a number of patients with mild but quite definite typhus fever in 1919 and subsequent years. He was struck by the number of these patients who were employed in grain stores or in transporting wheat, occupations likely to bring them into contact with rats and mice. Since none of the patients was louse-infected, he advanced the view that the disease was carried from rats or mice by some insect parasite. A few years later, in Queensland, there was an interesting relationship between a 'mouse plague', that is, a sudden enormous increase in the mouse population, and the appearance of numerous cases of typhus fever amongst farmers in the district. In both instances the connexion between rodents and typhus fever appeared to be unmistakable, but unfortunately no experimental proof that rats or mice were infected was obtained. The proof that typhus could be derived from natural infections amongst rats had to wait for American investigations in 1928. In the southern states cases similar to the Australian ones occurred, and it was shown conclusively that they were infected by rat fleas from rats chronically infected with the typhus rickettsia. Since then similar conditions have been found to exist in many other parts of the world. In all probability typhus is an ancient disease of rats and mice, perhaps an even more ancient disease of the fleas that live on the rodents. The typical louse-spread typhus is a modern development.

146

In Mexico it is possible to see the change from rat typhus to human typhus in progress. Rat typhus is common, and when conditions for case-to-case transfer by lice are present, a patient infected with typhus of rat origin by flea bite may set up a small epidemic of the classical type. There are some slight differences between the rickettsiae that produce the 'rat type' typhus and those of 'louse type' infections, but they are no more than the differences between local races of the same species. Zinsser, in his fascinating 'biography' of typhus fever, *Rats, Lice and History,* suggests the probability that the European louse-borne disease originated during the long-drawn-out wars of the sixteenth century in Hungary when Christians and Moslems alternately conquered and reconquered the country. No clear evidence of typhus fever is found prior to this period, but every war in Europe since has seen grave outbreaks of the disease. The greatest in all history was probably the epidemic which swept Russia during and after the Revolution of 1917, when cases were numbered in millions and deaths by hundreds of thousands.

The war epidemics of typhus have always been associated with famine, hardship and dislocation of social life. Under happier, if still crowded and filthy circumstances, typhus can take on a milder character. Following the general rule amongst infections, typhus is far milder in children than amongst adults, and in endemic regions such as the cities of North Africa a high degree of immunity may be developed by such mild childhood infections. Here we have reached the end of the process of transfer from rat to man. The mild disease of the rat passes accidentally to man, it finds a new vector in the louse, and under circumstances of war and famine spreads widely and fatally, but where circumstances allow its easy spread in a stationary population, it eventually develops the character of a typical relatively mild endemic infection.

These stories of human intrusion into rickettsial ecosystems illustrate the frequent finding that many of the most lethal infections of man are ecologically infections of other vertebrates – or of insects – which reach man only by accident. Other examples – plague, rabies, yellow fever, the haemorrhagic fevers – are discussed elsewhere in the book. One might even argue that *all* highly lethal infections of man, classical smallpox for instance, are caused by agents which have only recently (in a biological sense) become human diseases. One must be careful however not to suggest the converse, namely that *because* a human disease is mild it must represent an old-established equilibrium between the pathogen and man. Newcastle disease can be a dangerously lethal disease of domestic fowls, but in poultry handlers infected in the course of their work it produces no more than a short-lasting conjunctivitis.

A rather different type of evolutionary change seems to have been at work in Africa and concerns the diseases – and harmless infections – produced by trypanosomes, protozoal parasites of the blood which are transferred by the tsetse fly and were mentioned briefly in Chapter 4. The most important diseases they produce are nagana, a highly fatal disease of domestic animals which has played a big part in limiting the area of European occupation in Africa, and sleeping sickness of human beings.

Nagana is now known to be a complex of diseases that result from infection by one or more of at least six distinguishable types of trypanosomes. These include one form that in its appearance and life history appears to be identical with the *T. brucei* of sleeping sickness. All are transmitted by tsetse flies of the genus *Glossina*, and produce essentially harmless infections of wild ungulates. In general, where nagana is present human sleeping sickness does not occur, yet the trypanosomes obtained from sleeping sickness patients are identical in every respect with *T. brucei* from animal infections except that the animals strains will not infect human volunteers. All trypanosomes seem to be 'genetically labile' organisms rather readily developing resistance to chemotherapeutic drugs or changing virulence for laboratory animals, and there seems to be little doubt that the organisms of sleeping sickness are strains of *T. brucei* which have become specialized human parasites.

In the 'fly' regions tsetse flies are widely infected with *brucei* and other trypanosomes and any domestic animals or human beings entering the country are inevitably soon inoculated with the trypanosome. Horses, camels and pigs die rapidly, cattle more slowly, but equally certainly. Man is usually unaffected, even in heavily infected country where it is impossible to keep any domestic animals. In certain parts of Zambia cases were reported of rapidly fatal infections of human beings by a trypanosome of *brucei* type which were distinct from the chronic infections found in the sleeping sickness country to the north and west. It is a reasonable suggestion that such an occasional acute infection might represent the first stage in the adaptation of an animal strain to a new human host. Sleeping sickness is now ascribed solely to *T. brucei* strains which have been so long established in a man–tsetse fly cycle that they do not need an animal reservoir. Whether or not such a reservoir exists is technically impossible to determine with current techniques.

Current ideas about the evolution of the *T. brucei* group are that tsetse flies and trypanosomes have probably been present in Africa at least since the Oligocene period and a variety of tolerated host–parasite relations established with most or all mammals bitten by *Glossina*. This must have included man over the whole period of his evolution. From our present point of view the most interesting feature of the sleeping sickness situation is that this is a chronic human disease which, though eventually killing large numbers, maintains a constantly available reservoir of infection for the tsetse fly.

There are many other examples of such more or less balanced ecosystems to be observed in Africa. Those that interest us have nearly all been brought to light by the occurrence of fatal disease in domesticated animals introduced from abroad. There is no other part of the world in which large native mammals occur in such numbers, both of individuals and of species. Amongst them are zebras, antelopes and wild pigs, all close relatives of common domestic animals. Diseases have developed amongst these, and through centuries the various parasites and their hosts have become adapted to each other. A large proportion of these infections are spread by biting flies or mosquitoes, so that when domestic animals enter the country they can be infected without necessarily coming into close contact with the natural hosts of the parasites. It is only to be expected that domestic stock will be particularly susceptible to infections whose normal host is a closely related wild animal. If, to take an example, a virus producing a mild infection of zebras is transferred by a mosquito bite to a horse, it finds a host whose tissues are of almost the same character as the zebra's but lack the resistance which the latter species has developed to curb the virus. This may well be the origin of African horse-sickness, although actual proof has not yet been supplied. Amongst the diseases known to be derived from native animals are malignant catarrh, with the expressive Afrikaans name of *snotsiekte*, which is a natural disease of the gnu and affects sheep and cattle severely, and a highly fatal form of swine fever derived from a mild infection of the wart-hog. There are in Africa several other diseases of stock due to protozoa and viruses which must undoubtedly be derived from wild mammals. So far, however, the natural hosts have not been determined, an indication in itself that the natural diseases are inconspicuous.

Perhaps the most interesting of all such transfers of infection from one species to another is the infection of cattle with the virus of 'mad itch'. This is the name given to a rather rare disease of cattle in the mid-western states of America. The cattle develop an unbearable itching and almost tear their skin off against posts and the like for a day or two, then they develop signs of paralysis and invariably die. The virus was not difficult to isolate, and was found to produce fatal infections when inoculated in any of the common animals of the laboratory. Sheep and cattle were also highly susceptible to inoculation, but pigs were hardly affected, unless the virus was inoculated directly into the brain. Nevertheless, although pigs hardly ever died from inoculation, they became infectious for other pigs. A mild infection by the mad-itch virus passed from pig to pig in the experimental pens, spread apparently by the porcine equivalent of droplet infection. It could be shown that the virus was present in the nasal discharge of a pig sick with the disease, and that such discharge placed in a nostril of another pig resulted in infection. There was therefore a sharp distinction to be drawn between pigs and all the other animals tested. Monkeys, rabbits, guinea-pigs, sheep and cattle all died from inoculation of the virus, but they could not contract the disease naturally from one another. Pigs suffered a relatively mild but highly infectious disease. Although the virus had never been found naturally in American pigs, it seemed to Shope, who did all this work, that the indications pointed strongly towards the pig as the natural subject of the disease. It was known that after infection swine developed antibodies in their blood which could inactivate the virus. As in many similar problems, the easiest way to tell whether pigs were infected with the virus was not to look for the virus, but to see whether the pigs had developed antibody to it. Thirty or forty samples of blood from pigs killed in Iowa were tested, and all were found capable of rendering the virus harmless. Extending such work, Shope found that while pig farms in the eastern states were not infected, the majority of pigs in the middle west had passed through the infection apparently without any obvious symptoms.

Mad itch in cows only appears when cattle and swine occupy the same enclosures. Shope suggested that the common sequence that leads to infection is a cow resting on the ground; a pig infected with the virus nuzzles around, and its slobber infects either an existing scratch or a little abrasion made as the pig tries to push the cow out of the way. From the point of inoculation the virus spreads up nerves toward the brain. On its way it infects especially the ganglia of the nerves of sensation, and by its action on these nerve cells produces the intolerable itching around the point of entry of the virus.

In this example of an infection passing from its natural host to other species we have the same difference in virulence that characterizes the earlier examples mentioned, but an additional important point appears. The virus can spread easily only in the normal host species; infection occurs in other species only by abnormal and exceptional methods. It is much easier for a pathogenic microorganism to multiply in an alien host when it reaches its internal tissues directly than when it has to make its own way into the tissues. A highly specialized adaptation of virus to host cells seems to be necessary for simple contact infection to occur, using this term to include any form of infection of unbroken surfaces, either by droplet infection or by physical contact. It is particularly striking that of the virus diseases spread in this way hardly any can be caused to infect animals of another species by natural contact. Many are incapable of being transferred even by inoculation. Of human diseases in this group only influenza is infectious in the strict sense for any other animal, and here the susceptibility of the ferret must be regarded as something of a lucky accident. The same holds for animal infections spread in this fashion; rinderpest and foot-and-mouth disease in cattle, sheep-pox, and swine fever are all highly infectious, but with one exception are strictly limited in their infectiousness to their normal hosts. The one exception is rinderpest which in 1896 spread amongst the South African antelopes and buffalo and so depleted the game animals of the Zambesi and Limpopo basins that tsetse flies, formerly abundant, were unable to survive in the absence of their normal source of blood, and disappeared permanently from the region. Here is another example of the havoc a new infection can sometimes cause in species related to the natural host, but it is very much an exception to find such diseases spreading to the new host by the respiratory or enteric route. The disease spread by vectors which penetrate the skin with their bite are much more liable to infect unusual host species, and most of our examples of transfer between species have therefore been of this type.

*

A pathogenic microorganism can chance on one or other of an almost limitless numbers of means for ensuring its survival. In many ways one of the most 'satisfactory' from the parasite's point of view is to pass regularly from parent to offspring, necessarily persisting as a latent or relatively harmless infection until passage to the next generation becomes possible. Examples are rare but can exist at all levels. We have already mentioned transovarial passage of rickettsiae and spirochaetes in ticks and mites. There is an important disease of poultry usually transmitted through the egg and in somewhat similar fashion there is a mouse disease due to the virus of lymphocytic choriomeningitis (LCM) which passes regularly from the mother to the embryos in her uterus. It is a disease which has been widely investigated and has become of special interest to immunologists because of the curious tolerance of the congenitally infected mice to the disease. A heavily infected colony of white mice may appear superficially quite normal yet an injection of blood from one of them will kill any normal mouse from a clean stock.

There is no certain example of such intra-uterine or 'vertical' transfer from generation to generation in man but herpes simplex virus seems to have survived in what is ecologically a similar, if not strictly comparable fashion. It is transmitted 'vertically' in the sense that it can survive for generations in small human groups without the necessity of re-entry from outside.

Herpes simplex, cold sores on the lips, is to our knowledge the only virus disease mentioned by Shakespeare (*Romeo and Juliet* I, iv, 75) but we believe that it can be traced much further back, certainly to the common ancestor of man and the old world monkeys somewhere in the Oligocene and perhaps to the earliest mammals in the Cretaceous. More than thirty years ago one of us (FMB) was fascinated by herpes simplex and the ease with which virus and antibody could be assayed on the chorio-allantois of the chick embryo but almost equally by the social and evolutionary significance of the results.

A substantial proportion of people suffer from (or tolerate) recurrent herpes blisters, almost always on the lips. In a given individual the blisters tend to recur on the same area of skin on each occasion. They can be brought on by fever, hence the name 'fever blisters', or by a variety of other relatively trivial causes – a common cold ('cold sores'), sunburn, menstruation or an emotional upset. The blisters invariably heal without a scar within a few days. Fluid from a fresh blister contains much herpes virus and every person who suffers from herpes has antibody in his blood against the virus. Evidence on the point is scanty but what there is suggests that herpes infection persists through life in a latent form, becoming active on sporadic occasions, sometimes years apart. By no means every adult suffers from herpes or shows antibody in his blood and it is now clear that the primary infection is usually contracted between the ages of one and two years from some other member of the family, most often probably from the mother. In some infants the infection takes the form of extensive blistering and ulceration of the mouth; more often it is shown only by a trivial sore mouth. Observations of the habits of toddlers would suggest that hard objects easily contaminated by someone else's saliva are constantly being mouthed and little abrasions on the gums can well provide a portal for infection. Primary infection of areas of skin on the lips at this time presumably implants the virus which will wake into activity at intervals through life. There are some rare but serious, sometimes lethal, complications of herpes but in the present context we are concerned only with the basic method by which the virus passes from generation to generation. The sequence mother-to-infant makes it possible for the virus to survive under any conditions that allow survival of the host.

In Chapter 1 we indicated that a virus like measles could gain a foothold in a community only if there were a sufficient number of susceptible children available. In the larger cities the birth-rate is such as to ensure a continuing supply of non-immune children; as a result the common childhood diseases like measles, mumps and rubella remain permanently endemic. Comparison with the behaviour of the same diseases in island communities of various sizes has demonstrated that in any community of less than a few hundred thousand people measles will die out after an epidemic and will not reappear until it is imported from somewhere else. Measles, mumps and rubella obviously could not therefore have survived as diseases of primitive food-gatherers moving in small family groups. But if herpes simplex afflicted our Australopithecine ancestors of two millions years ago it could well have persisted to the present.

Monkey herpes is well known because at least a dozen people have died of it. Both the common monkeys used in research on polio, the Indian rhesus and the Malayan cynomolgus, carry a virus closely related to herpes which is as harmless to them as herpes is to us. For man, as a distantly related species with no evolutionary experience of present-day monkey herpes, the monkey virus is almost as lethal as rabies. There can be no doubt that the two herpesviruses, like their hosts, must have had a common ancestor. We may go even further back if we accept, as most virologists do, that the mad itch virus of pigs is another distantly related herpesvirus. To find a common ancestor for pig, man and monkey we have to go back to the very beginnings of mammalian evolution.

Before leaving herpes we must speak of a curious finding at the social level. In 1938–9 in Melbourne, 95 per cent of underprivileged children attending a public hospital had serological evidence of infection by herpes simplex, whereas those from comfortable middle-class homes showed a much lower proportion. Just eight years later Children's Hospital patients showed only 57 per cent positive. Rightly or wrongly we regarded and still regard this as an indication of the way knowledge of the value of care and cleanliness is spreading through the community. There could be worse indices of the level of civilization than the proportion of people who are *not* carriers of herpes!

There are three other viruses of the herpes group that infect man but from our present point of view only one need be mentioned – chickenpox (or varicella) and its *alter ego*, herpes zoster (or shingles). Chickenpox at first glance behaves just like measles; once a child has had it, he never gets it again. But when he is fifty or sixty or seventy he is liable to have an attack of shingles, which is a painful crop of blisters on the skin whose distribution corresponds to that of one of the sensory nerves, usually on the trunk. If there is a child in the house who has not had chickenpox that child will probably catch chickenpox from the shingles blisters. Shingles is in fact a recurrent infection by varicella virus that for forty of fifty years has apparently been resting in a latent condition in a spinal nerve ganglion. Here then is a possible way by which a virus can at least double its chances of survival. If it fails to be passed on from the child with chickenpox it may yet be transmitted years later from the grandparent with herpes zoster!

11. Control of infectious disease

It is natural that every community should endeavour to diminish the incidence of infectious disease within its boundaries. Enthusiasts have claimed that it needs only genuine international cooperation and con- centrated work to eliminate all serious infectious diseases from the globe. So much has been achieved in this direction even since the first edition of this book was being written in 1937, that something very close to this could now be formulated as a realistic and attainable goal. The advanced countries of the world have already eliminated every one of the plague diseases, plague itself, cholera, typhus, smallpox, malaria and yellow fever. Infantile dysentery, scarlet fever and diphtheria which were responsible for most of the childhood deaths in the nineteenth century are now rare and usually extremely mild. Some of these changes have occurred without any deliberate human action, but well-directed public health action has played a steadily increasing part. Today the principles of control for infectious disease are soundly based on fifty years' experience and can be applied almost as a routine.

Broadly speaking, the endeavour is to protect the *community* against sporadically occurring diseases by measures designed to prevent their entry and transmission and to protect *individuals* against endemic diseases. The first group includes several diseases which can be regarded as candidates for eradication from the world. In 1971 it is expedient to look rather carefully at the possibilities for world eradication. There is one good rule. If a country in which a disease was once endemic has rid itself of the disease and has shown that, if it is accidentally reintroduced, the focus of infection can be controlled and eliminated, then we can begin to think about eradicating that disease from the world. At the present time the World Health Organization (WHO) has proclaimed the eradica- tion of malaria and smallpox as realistic targets.

It is worth looking more closely at how the mode of transmission of an infectious disease influences the prospects of world-wide eradication as well as local control.

(1) Any disease that is transmitted only by the bite of an insect or other arthropod can be readily eliminated from any urban area. If the only reservoir of infection is the human one, elimination is possible in

Fig. 17. To show the progress of malaria eradication between about 1900 and 1960. The stippled area was once malarious but is now free, while countries in black still contain malarious regions. (Data from WHO. The question mark reflects the lack of data from China.)

Fig. 18. Distribution of smallpox in 1965: countries in which smallpox is still endemic are shown in black. (Data from WHO.)

any area which it is practical to cover completely. Malaria was removed with relatively little difficulty from heavily populated islands including Cyprus, Sardinia and Mauritius, all of which had been malarial for centuries. Where the main reservoir of infection is in wild mammals or birds, on the other hand, eradication is impossible unless the authorities are prepared to attempt the extermination of the reservoir, which may sometimes comprise several species.

(2) Any disease spread by the faecal–oral route which requires on the average a substantial dose to infect, e.g. amoebic and bacillary dysentery, cholera and typhoid fever, will not spread widely if there is normally effective sewage disposal and a reasonable standard of cleanliness. Eradication on a world scale is at present inconceivable. When the crowded and poverty-stricken countries of the world have brought populations down to a stable level and have raised their standards of living and community health to Western levels, the possibility may become worth considering. Experience in Europe and North America has however not been such as to encourage optimism in regard to eventual eradication. No affluent country has yet claimed to have eradicated typhoid, careful examinations anywhere will show people who carry enteric amoebae that look exactly like the pathogenic ones, and disconcerting outbreaks of bacterial dysentery occur occasionally. We may well have to be content with maintaining a degree of control of intestinal infections that will always be far short of eradication. The difficulties are naturally even greater with the intestinal infections due to viruses, of which infectious hepatitis is now the most important. With viruses the infective dose is small enough to be transferred indirectly via fingers and minimally contaminated objects.

(3) Infections spread by the respiratory (droplet) route can in practice only be blocked by immunization. If, as in the case of smallpox or measles, the disease is caused by a single antigenic type of virus and firm immunity follows infection or immunization, the chances of eradication by immunization are good. In fact only with such viruses can we expect good immunity after vaccination. For any disease of this type to spread, a certain minimum concentration of susceptible hosts must be available. If by universal immunization in infancy this is never allowed to develop, and there is no way in which the virus can survive for years as a latent infection in a carrier, the virus must disappear. Mumps and rubella might also be eradicated in this way but there are more urgent public health problems to be dealt with before that is attempted.

Characteristics of an infectious disease that virtually rule out the possibility of world-wide eradication can also be listed briefly:

158

(1) Infections of wild animals and birds where human infection results from intrusion into an alien ecosystem, e.g. jungle yellow fever, scrub typhus, psittacosis.

(2) Infections that persist throughout the life of the carrier and are transmissible to others years after they were originally contracted. Fortunately the two common human examples, herpes simplex and varicella–herpes zoster, are of minor importance.

(3) Infections spread by the respiratory route in which the disease in question may be produced by viruses of many different antigenic types. Certainly in influenza and probably in regard to the common cold viruses, such agents have also a high capacity to change antigenic character by mutation. This phenomenon of antigenic drift will be discussed in the chapter on influenza.

(4) Infections in which it is socially impossible to obtain public cooperation. We have the venereal diseases particularly in mind, but in some ways this will also apply to any type of infection that is regarded as not serious enough to justify the effort involved.

With this background we can describe a few of the more interesting and important successes in the control of infectious disease. The examples have been chosen to illustrate the three basic approaches that are available to the public health authorities. They are: first, preventing the entry of the parasite by quarantine measures, secondly, interruption of the chain of transmission by what can be broadly called environmental sanitation and, thirdly, protection of the susceptible individual by immunization or chemoprophylaxis. It is perhaps indicative of the forces that drive men to effective action that most of the examples to be cited have been spurred by war or the urgent needs of commerce.

*

Quarantine for infectious disease has been mentioned briefly in Chapter 9 largely in the context of its importance for Australia. It is clearly expedient that when a country is free of a disease every effort should be made to prevent its entry. Quarantine originally meant holding travellers arriving from an infected region in some isolated quarantine station for long enough to ensure that if any of them were in the incubation period of the disease they would develop symptoms. If all remained healthy they were then released as presenting no danger to the community.

The approach has been progressively modified and with the virtual disappearance of the passenger liner, quarantine is now hardly ever applied. Instead it is insisted that all travellers from countries where smallpox or yellow fever may be present must have a current certificate of vaccination against the appropriate virus. The only other common routine is employed when passengers arrive from a region known to be currently subject to some significant outbreak of disease. Each individual is given a card instructing him to show it to any doctor whom he has cause to consult within the next two or three weeks.

Much more drastic measures are applied to prevent the entry of animal disease, often to the extent of completely forbidding the entry of some species of domestic animals. New difficulties have arisen in regard to the air transport of frozen bull semen or the shipment of fertile eggs from one country to another and most countries have developed rules and procedures which they believe will cover any risks of bringing in disease in this fashion.

Animal disease is of such immense economic importance that if ever quarantine control breaks down and an outbreak occurs, extreme measures to contain the outbreak are invariably applied. In Australia rinderpest of cattle and Newcastle disease of fowls have both been successfully eliminated after accidental entry into the country.

*

Where the need is to block the chain of transmission between source and susceptible, the means required will vary greatly according to circumstance. Australian cities, like those of most Western countries, are free from bubonic plague but in the past the disease has entered seaports and established itself in city rats. When this happened the remedy was to institute active rat-catching and poisoning measures, with regular bacteriological examinations, and to educate the public about the danger of rats and the necessity for grain and food stores to be made adequately rat-proof. So far, such action has been sufficient, although in the case of the New South Wales outbreak it took ten years to eliminate all plague-infected rats.

In Australia the 'gold rushes' always resulted in serious outbreaks of typhoid as a natural consequence of the extremely primitive sanitary arrangements which were inevitable in the early 'canvas town' days of a new goldfield. Nowadays, in addition to organized removal of excreta either by water carriage or, in smaller towns, by the most satisfactory method within the economic ability of the community, control is exercised over the sanitation of temporary or permanent camps involving more than a small number of habitations. A clean water supply and the prevention of known carriers of the typhoid bacillus being employed in any trade involving the handling of food, complete the defences of the community against typhoid.

A classical example of the approach in dealing with serious spread of infection in a socially disorganized community can be found in the story of how typhus was dealt with when Naples was occupied by allied troops in December 1943. It was the first great demonstration of the large-scale use of DDT as a 'delousing' measure. Naples had been heavily bombed and a large proportion of the population had been living under crowded and insanitary conditions in caves and other types of air-raid shelter. Conditions for the spread of lice (and the typhus rickettsia) were ideal. The problem was to free half a million people of lice within the shortest possible time. The old methods of providing a bath and local treatment for each individual while his clothes were being heat sterilized would have been physically impossible in the time available. Fortunately new techniques for handling DDT to the best advantage had just been developed. With proper equipment a mixture of DDT powder and an inert dust could be blown into the clothes in a matter of seconds and, in the great majority of instances, all body lice would be killed in an hour or two. The American army personnel applying the method were immunized against typhus. Some tons of DDT dusting powder and the necessary blowers were available and for the first time in history a menacing typhus epidemic was abruptly terminated.

Another example of this type of approach was the cleaning up of the Panama Canal Zone. Malaria and yellow fever, both mosquito-borne diseases, were the chief hindrance to the construction of the canal. By intense attack on the breeding-places of mosquitoes, by universal screening of houses and by the isolation and treatment of all malaria and yellow fever patients in mosquito-proof hospitals, both diseases were practically eliminated. Environmental control for the prevention of malaria has now become a highly technical matter requiring a detailed knowledge of mosquito ecology and human cultural patterns in each affected area. Examples of the problems and how they are overcome can best be given in Chapter 18.

As a final example in this group we can mention a minor disease which is just not of sufficient importance in the public eye to justify the measure which would eliminate it. This is psittacosis. One has only to eliminate the trade in cage birds to get rid of the disease in man. Alternatively a close control of the commercial breeding of budgerigars, the most important reservoir, was for a period just before the Second World War shown to be effective in California. But nobody was sufficiently interested and as far as we are aware there is no positive action anywhere to prevent psittacosis at the present time.

*

When we come to the methods of control which are based on protecting the individual rather than the community we find two distinct approaches. We may wish to protect the whole population and the only practical method will be by universal immunization in childhood. The second approach is called for when persons from a non-endemic area must work or fight in an area where the disease in question is perpetually present, or where an individual is exposed for any reason to specially high risk of infection. Under such circumstances various methods may be used. When protection from mosquitoes is impossible, prophylactic doses of an appropriate drug will prevent any symptoms of malaria. Venereal disease can be prevented by those who will take the necessary precautions after exposure to infection. It will usually be possible to avoid cholera in the most heavily infected region if only cooked food and boiled water are taken. Other examples will suggest themselves, all on one of two principles, either to take some extra precaution to prevent infection reaching its usual point of entry into the body, or to apply some medicinal remedy which is capable of dealing with the infection before it develops far enough to give symptoms.

Since potent drugs became available the method of chemoprophylaxis has been widely tested. The use of quinine to prevent the appearance of bouts of malaria had been common practice for many years when early in the Second World War the main supplier of quinine (Java) became no longer accessible and mepacrine (Atebrin) replaced it as the standard prophylactic for American and Australian troops in the theatres of the Pacific campaigns. Later proguanil (Paludrine) and chloroquine in turn replaced mepacrine, and these have remained standard except in regions such as South-East Asia where during the last two years resistant strains of malarial parasite have appeared and quinine is returning to favour.

These chemoprophylactic measures against malaria provide almost the only example of mass chemoprophylaxis still used routinely. At one time there seemed a possibility that penicillin or an appropriate sulphonamide could be used to prevent all the common bacterial infections. However the virtual certainty that the dominant strains of bacteria become resistant to any initially effective drug was soon realized. Clinical trials that were made encountered serious difficulties and they are better discussed in the chapter specifically concerned with the antibacterial drugs. In general, chemoprophylaxis can be used effectively only for the protection of individuals who are known to have been exposed to the danger of infection by some specific episode. The pathologist who pricks his finger while doing a post-mortem examination, the patient who is having a major operation on his lung, the doctor who has examined a patient with pneumonic plague, or the sailor who has exposed himself to the danger of venereal disease – any of these can be protected by appropriate treatment with one or other of the anti-bacterial drugs. Widespread protection against the possibility of infection over long periods is not a practical proposition.

In standard medical practice today the most important use of chemoprophylaxis is in protection against recurrent rheumatic fever. Children with rheumatic fever almost regularly suffer some degree of heart damage. The damage in both joints and heart is an indirect result of streptococcal throat infection, and subsequent streptococcal episodes over the years are liable to increase the cardiac damage. It is necessary therefore to do everything possible to protect the children against streptococcal infection and most such children are now taking penicillin tablets every day of their lives or at least until they are fully adult.

*

Throughout this book we have emphasized that most infections do not lead to disease but merely to a subclinical infection which confers immunity just as surely as the disease itself. The logical aim of public health measures should be to make it possible for all to gain this type of immunity without risk of overt disease. It is the most important task of childhood to do so, because, as we have seen, childhood from the ages of two to fifteen is the period at which infections first met are likely to be least dangerous. Unfortunately, we cannot as yet say what determines whether or not an infection takes a clinical or a subclinical form. Common sense, plus a little, but only a little, relevant evidence, suggests that infections are most liable to be subclinical in children who are well nourished and healthy at the time of infection and who receive only a small dose of the infecting agent. Such a point of view provides us with certain practical rules for the upbringing of children. From an early age all children should mix freely with others, but in the open air or in well-ventilated school rooms. The aim to be kept in mind is that in the period from six to twelve years children should be infected with and overcome without damage each of the common endemic infections, meeting them in small amount and as far as possible at times when the infective agents are not of abnormally high virulence. Unless this is done in childhood the individual will reach adult life without immunity, and will be more liable to be severely infected when he eventually meets the disease.

The development of the various vaccines that we now use to protect children represents basically our response to the need to provide immunity for all without the risks associated with natural infection. The same or similar vaccines are even more urgently called for when adults from one country have to live in another where a disease which they have not previously met is endemic. Actually the greatest stimulus to develop methods of immunization has been military necessity. When European troops were needed in countries where infections like typhoid fever and cholera were endemic, it was necessary to find some means of protecting them.

When it was believed that the development and persistence of antibodies in the blood were wholly responsible for recovery from disease and subsequent immunity, the requirements for prophylaxis seemed to be obvious. What was needed was to provoke the body by some harmless method to produce the proper type and amount of antibody. Even with a more sophisticated view of the complexity of the immune response, it should still be possible to induce the essential features by something less than actual disease. In the most general terms the way in which this is attempted is to prepare antigenic material from the corresponding microorganisms in some non-infectious and non-poisonous form and inject this under the skin. There are two difficulties: on the one hand we must avoid using anything so like the causative microorganism as to be dangerous to the person inoculated, and on the other we must be sure that in preparing the artificial antigen we retain all the qualities which are needed to produce an effective response. It is not infrequent to find that when a bacterium or virus is killed by heat or antiseptics it loses just those antigenic qualities which are required to produce real immunity.

The most effective but also the most dangerous way to produce immunity is to inoculate with living organisms whose virulence has been reduced to below the danger level by some suitable method. The very first method used to protect children against smallpox was by the inoculation of virus of undiminished virulence in the form of matter from a natural smallpox pustule on another child. Actually this was probably not quite so dangerous as it sounds, and the results obtained were good enough to make the method relatively popular for fifty years. The inoculation was made into the skin, and an infection starting here was much less likely to be severe than one contracted naturally by inhalation of the virus into the nose or throat. This 'variolation', as it was called, was eventually

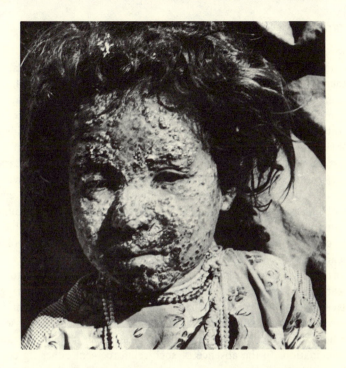

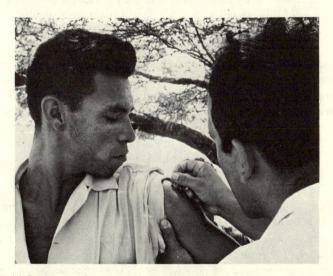

Fig. 19. (a) Smallpox: a five-year-old Afghan girl whose face will be pitted for life. (b) Smallpox: the simple answer, Jennerian vaccination. (Both plates by courtesy of WHO.)

replaced by Jenner's method of inoculation with cowpox – 'vaccination'. The history of vaccination is full of controversy and misunderstanding; neither Jenner and his supporters nor his opponents showed much of the scientific attitude in their endeavours to prove on the one hand that vaccination was an absolutely certain protection against smallpox, or on the other that it was harmful and dangerous. There is even considerable doubt as to what sort of virus was actually used for vaccination in the years immediately following Jenner's discovery. Recent comparative studies of the viruses of smallpox and cowpox with the standard vaccine virus now used for vaccination show that vaccine virus more closely resembles smallpox than cowpox. We can feel reasonably certain that vaccinia virus is a lineal descendant of true smallpox virus which has lost most of its virulence and contagiousness, but is still able to provoke the appearance of the same type of antibody.

Vaccination was probably the most important reason for the disappearance of smallpox as an endemic disease from Europe during the nineteenth century. When smallpox was endemic, a child vaccinated in infancy would probably come into contact with smallpox on several occasions during childhood and reinforce the basic immunity acquired by vaccination. In the absence of such reinforcement either by contact with the disease or by revaccination, immunity fades and vaccination in infancy is of little significance when first contact with smallpox occurs thirty or forty years later. There are bound to be differences of opinion as to the most desirable policy in the countries where smallpox is not now endemic. All require that any traveller from an area where the disease is endemic should have valid evidence of vaccination within the last three years. Some still insist on universal vaccination. In England and Australia the preferred approach is to make vaccination compulsory only as an emergency measure when importation of a case occurs and to use this as a barrier to spread in threatened localities. In endemic areas of the tropics and subtropics vaccination is unequivocally necessary for all.

Vaccination against smallpox can be taken as typical of those methods by which living but only slightly virulent organisms are inoculated into the body. In recent years the use of such methods has spread rapidly, particularly for protection against virus diseases. In addition to anti-smallpox vaccine there are now well established vaccines against yellow fever, polio (Sabin vaccine), measles, rubella and mumps. Living vaccines which are not yet at the stage of general use have been prepared against rabies and influenza.

In 1971 the Sabin live virus vaccine against polio is widely used in almost every part of the world. It is perfectly safe in children and has eliminated paralytic polio in every country which has been completely immunized. For a time there was much discussion of the possibility that the virus might revert to virulence if it passed from one individual to another. This was something that no theoretical discussion could settle but ten years without any suspicion of such an occurrence has given the Sabin vaccine a guarantee of stability and safety. Like other living virus vaccines it provokes an infection basically similar to that of a virulent strain but in a lower key. It is administered by mouth as a few drops of palatable syrup on a disposable spoon – something highly acceptable to apprehensive young children with a fear of needles! When the vaccine virus reaches the intestine it infects the cells of the bowel wall, concentrating probably on the lymphoid tissue accumulations we call Peyer's patches. Here, and in the abdominal lymph glands and spleen, antibody is produced and new lines of lymphoid cells that can react sharply to renewed contact with the virus, laid down. In one way or another, most probably by leakage of antibody and lymphocytes into the bowel, the intestine develops a local resistance to any fresh implantation of virus, a resistance which is not produced by Salk vaccination.

There is abundant evidence that the method is effective in preventing paralysis but not that it is more effective than the standard three doses of Salk vaccine. Theoretically one would expect that Sabin vaccination should be more effective than Salk in preventing any natural reimplantation of virus in the bowel. This is the case soon after primary vaccination but it is not known how long effective immunity lasts in the absence of casual reinfection. There is a possibility in principle that polio could be eradicated from the world by universal immunization followed by regular booster doses for a decade or two. Human nature, however, as well as the virtues of the vaccine must come into the calculation. In ostensibly enlightened countries there is already difficulty in persuading parents of the importance of primary immunization with Sabin vaccine when paralytic polio cases are no longer occurring. Multiplying those difficulties on a world scale makes us pessimistic about eradication.

Measles live virus vaccine has been one of the leading new developments in public health in the 1960s. It was developed by 'adapting' the virus to multiply in cultures of chick embryo cells. The main difficulty was to bring the virus to just the right level of virulence and stabilize it at that level. The first practical vaccine using Enders' Edmonston strain of virus was highly effective but with just a little too much bite. It was injected under the skin and in seven to ten days' time the child had a minor rise of temperature and usually a faint rash if one looked carefully enough, but rarely any other symptoms. This was the picture in North American children but when the vaccine was tested in Africa some of the children reacted with what was almost a typical attack of measles. There were two ways to deal effectively with this difficulty. One was to give a dose of human serum containing measles antibody with the vaccine. The other was to attenuate the vaccine virus further, though not so far that it failed to produce adequate immunity. In the last few years there has been a striking reduction in the recorded number of measles cases in America such that measles, the best-known disease of children, has almost

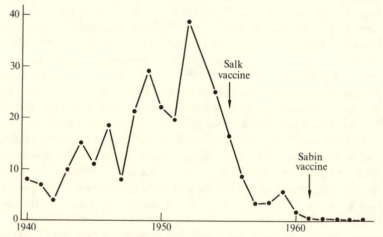

Fig. 20. Poliomyelitis in the United States, 1940–65. Annual incidence per 100,000 population. (Redrawn from J. R. Paul, 1971.)

disappeared from some of the larger US cities. Both the Schwarz and 'Moraten' attenuated virus vaccines appear to be equally effective.

At one stage paediatricians were sceptical about the need for measles immunization. Measles was one of the standard episodes of childhood that rarely seemed to do any harm now that antibiotics were available and it was felt that it was better to acquire a firm lifelong immunity from a real attack than partial immunity of unknown duration from a vaccine. The swing of opinion was mainly due to a growing realization of the frequency of damage to the brain during a measles attack. Measles encephalitis is rare, the estimates of its frequency ranging from one in every thousand cases of measles when minor degrees of cerebral involvement are accepted, to one in 10,000 when the criterion for inclusion is frank clinical encephalitis. There is a high mortality in these severe cases and most of the survivors show evidence of disabling brain damage. Careful observation of uncomplicated measles will often show signs of minor brain involvement as evidenced by an altered electro-encephalogram and there have been suggestions that some of the personality and learning difficulties in children result from measles. So in America at least the objective is now defined to have all children immunized against measles in their first two years of life. So far nothing has emerged to suggest that there are any hidden dangers of universal measles vaccination and the immunity is lasting well.

Optimists have pointed out that if, as seems almost certain, any person immunized against measles cannot subsequently become a source of measles infection for others, measles could in principle be eradicated from the world by general immunization. Even if only 90 per cent of children were immunized this would probably ensure that measles would fail to spread. Measles seems to be caused by exactly the same virus everywhere in the world, and if this holds true, ten years of universal vaccination might get rid of it completely.

Amongst bacterial diseases, living attenuated cultures have been used for immunization against plague with considerable success. In a later chapter we shall mention the BCG vaccine against tuberculosis, now being very widely used but still not universally acceptable to epidemiologists.

The use of living organisms for immunization on a large scale is always liable to occasional disasters. It may be possible to be quite certain that the organism we intend to inoculate is harmless, but there is always a possibility of some unwanted and harmful microorganism finding its way into the material. Cultures or tissue extracts containing living microorganisms to be used for inoculation cannot be sterilized, so there is no automatic method by which their safety can be assured. In any reputable laboratory every safeguard will be taken to eliminate any possibility of danger from contamination, but all these precautions depend ultimately on human vigilance, which is never infallible. There was a strong prejudice in favour of using immunizing preparations which could be sterilized by some reliable method and therefore were necessarily free from any living microorganisms. History, however, suggests that accidents can happen with either type. On the one hand with live vaccines we have the Lubeck disaster with BCG and the enormous number of American servicemen who developed hepatitis after yellow fever immunization in 1941–2. Killed vaccines produced the 'Cutter episode' when virulent polio virus was somehow present in a Salk vaccine and the Bundaberg tragedy referred to earlier where diphtheria toxin–antitoxin became contaminated with *Staphylococci*. Whatever type of vaccine is to be used, it must be prepared and administered with care and precautions that will cover extraordinary accidents as well as routine dangers.

Most killed vaccines in common use are suspensions of killed bacteria in some suitable antiseptic fluid. Vaccines against typhoid and para-typhoid fever and cholera are still in general use, but few people now regard such vaccination as an effective substitute for environmental hygiene. When the British Army was in occupation of the Canal Zone in the postwar years, typhoid fever still produced cases in troops properly vaccinated with what was then regarded as the best typhoid vaccine in the world.

A notable development during the 1939–45 war was the use of a typhus vaccine prepared with killed rickettsia that had been grown in chick embryos. This vaccine plus the use of the new insecticides kept the Allied troops in North Africa almost wholly free from the typhus which at the time was widespread among the civilian populations.

The best known of the killed virus vaccines is the Salk vaccine against polio which has been discussed earlier. This is essentially a suspension of the three types of polio virus rendered non-infectious by treatment with formalin. The Salk vaccine was of tremendous importance in showing that immunization against polio was possible but it has now been replaced almost wholly by Sabin-type live virus vaccines.

In civilian life the only bacterial disease which calls for prevention by vaccines of this type is whooping cough. It took many years of work to produce a really effective vaccine, but there is no doubt that modern vaccines confer a high degree of resistance for several years. Un-fortunately it is moderately toxic and very rarely a child may suffer severely as a result of vaccination.

The final type of immunization to be mentioned is that directed against infections which produce symptoms by the liberation of bacterial toxins. Diphtheria is the best-known example but tetanus (lockjaw) which results from infection of wounds with tetanus bacilli is today at least as important. As we have already discussed, immunity to these infections is almost entirely a matter of antitoxic immunity. In the absence of their toxins the bacteria would be harmless. Artificial immunization is therefore logically directed toward provoking the appearance of antitoxin; antibodies against the bacteria themselves are not required. Methods of preparing the toxins of these bacteria have been known for many years, but the toxins themselves are too poisonous to be injected as such. Fortunately there is a simple method by which any bacterial toxin can be rendered non-poisonous, but still capable of provoking the formation of antitoxin. Treatment with formalin will convert the toxins of diphtheria or tetanus bacilli into 'toxoids' with these desirable qualities. The use of tetanus toxoid has been standard military practice for many years and in view of the fact that tetanus, though rare, is also important in civilian life, orthodox opinion is that immunization against tetanus should be a standard procedure. The current practice is to combine diphtheria and tetanus toxoids with whooping cough vaccine as a 'triple antigen' for immunization in infancy.

*

At the end of any discussion of the control of infectious disease one thing must be emphasized – that control measures cannot be abandoned until the causative organism has been eradicated from the whole world. As advanced nations progressively eliminate diseases like polio and measles as well as smallpox, there will be a strong temptation to abandon the costly and troublesome procedures which made that consummation possible. The man in the street, and perhaps even governments, may become blasé about a disease that nobody has seen for years and seems to pose no obvious threat. But rapid intercontinental air travel has transformed the world into an epidemiological unit and, as long as a single pocket of endemic infection remains anywhere on earth, the possibility is real that the disease in question will one day be reintroduced into a country from which it has been eradicated.

Such a possibility has to be considered with special care in any plans for the local eradication of malaria. As long as an area from which malaria has been eliminated adjoins an uncontrolled area, constant vigilance will be needed to maintain a satisfactory control. If, for instance, all sanitary control were withdrawn from Panama, we should probably see malaria rampant again within a few years. In the population who had grown up in the region without experience of infection and with no immunity the intensity and mortality of malaria would almost certainly exceed that of a district which had never been controlled.

This is the great limiting factor in such attacks on endemic disease. Unless a disease can be exterminated from the whole globe, the precautions necessary to prevent its spread must be permanently maintained and incorporated into the social life of the community. If a population has been freed from any disease still prevalent in a country with which it communicates, it is in a peculiarly dangerous position should control ever break down. In the absence of constant immunization by clinical or subclinical infections, a highly susceptible population will arise, and if the infection reenters at a time when the measures that eliminated it have become ineffective, the resulting epidemic will be abnormally severe.

We can applaud the decision of WHO to work for the eradication of malaria and smallpox from all countries. But we do not envy the man whose responsibility it will be to nominate the day when smallpox is deemed to have been finally eradicated so that vaccination may be abandoned for all time.

12. Antibiotics

Until the early 1930s it was taught and believed that whereas protozoa and spirochaetes were susceptible to drug treatment bacteria were for some obscure reason outside the reach of chemotherapy. Only immunological approaches were regarded as worth following. It was natural for instance that the intense scientific interest at that period in the pneumococcus and pneumonia should have led to vigorous efforts to produce an effective antipneumococcal serum, and some success was being obtained. It was thought that in due course similar methods might be applied to other bacterial diseases. Then in 1935 came the beginning of a therapeutic revolution with Domagk's discovery of the effectiveness of prontosil against streptococci. Domagk discovered prontosil, the first sulpha drug, by following an idea that was a guiding principle for Ehrlich, that dyes stain living cells by much the same process as certain drugs kill them. Prontosil was a red dye that killed streptococci, but it was soon found that the effective part of the compound had nothing to do with its colour. The sulphonamides that followed were all derivatives of sulphanilamide, which to a near approximation is the uncoloured half of the prontosil molecule. The discovery that some bacterial diseases at least were susceptible to chemotherapy precipitated a search for new antibacterial agents. Penicillin was the next and greatest achievement.

Penicillin came into practical use in medicine by a rather roundabout route. In 1922, Fleming saw a mould colony on a bacteriological plate that appeared to be dissolving adjacent colonies of staphylococci. Following up this chance observation he showed that this particular mould, a species of *Penicillium,* produced something which diffused into the surrounding fluids and was very effective in preventing the growth of certain bacteria. At the time, however, only half-hearted attempts were made to use this substance for the treatment of human infections. For several years it found a technical use in bacteriological laboratories as an aid to the isolation of bacteria like the whooping-cough bacillus which are insensitive to its action. Most bacteria in the mouth are incapable of growing in the presence of penicillin, and the standard method for confirming a diagnosis of whooping cough depends on the use of a plate

173

of nutrient medium containing an appropriate amount of penicillin. Obviously it is easier to see small colonies of the whooping cough bacillus if the ubiquitous staphylococci and streptococci of the mouth and throat are prevented from growing.

An important stimulus to the further study of penicillin was probably provided by Dubos's work in America. He was a French bacteriologist who conceived the idea of using bacteria from the soil to destroy disease-producing bacteria. His early training had been as an agricultural chemist and he had an intimate knowledge of the enormous variety of bacteria in ordinary soil. By a special technique he isolated a bacterium from soil dug from a cranberry bog which produced something that destroyed pneumococci. Then he went on to isolate the substances responsible, cyclic polypeptides chemically, which he called gramicidin and tyrocidine. These were extremely active in killing pneumococci and other bacteria in the test tube, but they were too toxic to be of much use in treating human disease. Dubos's experiments created great interest among bacteriologists, and suggested that other products from microorganisms might be even more effective. At the beginning of the 1939 war, Florey at Oxford was looking for a project that might offer a significant contribution to the war effort. He knew that in addition to Dubos's substances, anti-bacterial agents had been observed in cultures of Fleming's mould and of the bacterium *Pseudomonas aeruginosa*. He almost literally tossed up to decide on which to study, and chose penicillin with the result that is known to everyone. A method of producing and purifying penicillin was worked out and its effectiveness in treatment demonstrated. The technical problems of large-scale production were sent across the Atlantic for solution, and penicillin in quantity was ready to play its part in the liberation of Europe.

Once penicillin had proved its quality a furious search began for similar products of fungi and bacteria which might have a wider or different range of action. Most of the significant discoveries have been made by the American drug firms or by workers closely associated with them. Large numbers of these 'antibiotic' drugs have come into use in medicine and there are hundreds more which for one reason or another have been judged unsuitable for practical use. Virtually all came from soil organisms – some from fungi, others from bacteria, but most from *Streptomycetes,* a strange sort of filamentous bacterium. The ecological significance of the antibiotics secreted by these organisms in their natural environment would doubtless make a fascinating study in its own right, but almost nothing is known about the subject. It is possible that their production confers a survival advantage by killing adjacent micro-organisms but there is no positive evidence for this. All the antibiotics come from spore-producing organisms and there have been suggestions that they may have a biological function concerned in some way with cell wall development and sporulation.

The discovery of a useful new antibiotic has always been the result of a tedious empirical screening of microorganisms of every origin for antidacterial properties. As soon as a clinically successful one emerged, however, biochemists and bacterial physiologists found much to interest them in defining the structure and probable mode of action of the agent.

The word 'antibiotic' was probably coined to parallel 'antiseptic' and mean an antibacterial agent produced naturally by some type of lowly organism. It was some years before the chemical composition of the early antibiotics was established but this is now known for all the important examples. With the development of synthetic methods for producing the antibiotic chloramphenicol, and semi-synthetic methods for producing modified penicillins the name has virtually lost its original significance. It is immaterial today whether an antibacterial drug is a natural product or wholly synthetic and there is a growing tendency to use the term 'antibiotic' to embrace all antimicrobial chemotherapeutic agents. In practice all that matters is that to be useful in medicine the antibacterial agent must be wholly or relatively harmless to human cells. It is this harmlessness that differentiates penicillin and the like from the antiseptics and disinfectants of the Listerian era. Carbolic acid (phenol), bleaching powder (hydrochlorite) and formalin are all potent in killing bacteria but they are almost equally damaging to body cells and could not conceivably be injected into the body, though they have their role in disinfecting inanimate objects.

A chemotherapeutic agent must have a 'selective toxicity' for bacteria, which in essence means that it can block or interfere with a metabolic pathway, or a particular enzyme concerned in such a pathway, which is vital to the bacterium but which is either not found in or is unimportant for human cells. Penicillin, for instance, specifically inhibits the synthesis of the cell wall in those bacteria sensitive to its action. The cell wall, as was mentioned earlier, is a very important part of the bacterium and only under special highly abnormal conditions can a bacterium remain alive without its rigid support. The bacterial wall contains several components not found in any animal body, including abnormal amino acids, right-handed in structure instead of left-handed, and muramic acid, a complex carbohydrate. Penicillin interferes with the process by which these and other substances are built up into the complex glycopeptide structures of

Natural history of infectious disease

the bacterial cell wall. It is because there is nothing remotely resembling such a structure in human cells that penicillin is so completely non-toxic for man.

One of the most interesting aspects of penicillin to those who worked with it in the early days was that it had no action on a bacterium that was not growing; only when growth commenced did the organism become susceptible. The limitation of its action to the cell wall supplies the answer. When a susceptible bacterium begins to grow in the presence of penicillin the cytoplasm expands while the cell wall remains unchanged. Eventually the organism literally bursts out of its shell to form a spherical 'spheroplast'. Without the protection of the rigid cell wall the spheroplast is unable to handle the osmotic pressure generated in its protoplasm and explodes. The cephalosporins also hinder cell wall synthesis but other antibiotics work differently. For example, chloramphenicol, the tetracyclines, and the so-called aminoglycosides like streptomycin and neomycin interfere with bacterial protein synthesis at one or other step in the process. Protein synthesis in mammalian cells conforms to the same general pattern but differs sufficiently in detail to allow the cells to withstand any potentially damaging effect of these antibiotics. A good antibiotic must be able to capitalize on such relatively minor differences in the metabolic processes of bacterial and animal cells. This has an important bearing on the development of 'resistant strains' of bacteria.

*

As we have discussed in Chapter 3 one of the outstanding qualities of bacteria is their capacity to change, either by mutation or by the incorporation of genetic information from another bacterial strain. Antibiotics provide such an intensely selective environment that any change which brings about resistance while still leaving the other functions needed for infection intact will mean that the new form rapidly replaces the old.

The story of what has happened to the common staphylococci since the advent of penicillin is enlightening. Nowadays it is rare for staphylococcal infections arising in hospitals to be due to organisms sensitive to penicillin. Most of them are due to staphylococci which can produce an enzyme destroying penicillin and so are relatively insusceptible to its action. It is something of a mystery where these organisms come from. In all probability they have always been present as a small proportion of the population of staphylococci but only found an opportunity to flourish when the environment in which a hospital staphylococcus had to live

176

contained penicillin almost as a matter of course. A staphylococcus that produces the enzyme penicillinase has an automatic way of ensuring its survival against any concentration of penicillin likely to be present in the body. If any such are present in the environment, and if they have any invasive capacity, they will soon come to dominate the situation in an infected wound. Once penicillinase producers are common in a hospital, transfer of the penicillinase gene to other and perhaps more virulent staphylococci becomes a possibility. This has been demonstrated to occur in the laboratory by carriage in an appropriate bacterial virus and there are other possible modes of gene transfer which may have helped to increase the frequency of penicillin resistance in hospital staphylococci. Mutant strains of staphylococci that are resistant to penicillin for other reasons can be produced experimentally but they appear to be of no clinical importance.

The ideal way of dealing with the situation was clearly to find or synthesize a penicillin which was not destroyed by penicillinase, and in 1960 British chemists found a way to do this. Penicillin is not a very large molecule but its construction is complex and subtle. Its formula is known and very small amounts have been synthesized by a rather inefficient process that could never be a commercial proposition. Short of complete synthesis, however, there are two ways by which penicillin can be modified. As produced in culture, 'penicillin' is a mixture of substances with a common chemical nucleus (whose name can be abbreviated to 6–APA) with one or other of several possible side-chains. By incorporating appropriate chemicals like phenylacetic acid in the culture fluid we can ensure that the great bulk of the penicillin produced has that particular side-chain which experience has shown to give the most active penicillin. When a medium containing no chemical components that can provide an effective side-chain is used, the mould produces 6–APA itself, which in that form has no antibacterial action at all. It was found in 1960 that this nucleus could then be united to chemical groups of many different types. The activity and usefulness of the resulting 'semi-synthetic' penicillins depend on the nature of this side-chain. Some, such as methicillin and cloxacillin, are resistant to penicillinase; others, such as ampicillin and carbenicillin, are active against problem bacteria like the typhoid bacillus and *Pseudomonas* that are insusceptible to the original penicillin.

Penicillin has remained the best and most widely used of the antibiotics but its effectiveness depends on the susceptibility of the infecting bacterium. This holds equally for every other antibacterial drug. Serious streptococcal infections have almost vanished everywhere simply because all virulent streptococci are sensitive to penicillin and no resistant forms have emerged. Staphylococci, on the other hand, have proved adept in developing any type of resistance that is needed for their survival in a hospital environment. By 1950 most 'hospital staphs' had already become resistant to penicillin by virtue of their penicillinase, but following the introduction of other antibiotics they proved equally able to develop resistance to all of them too, including most recently even the penicillinase-resistant penicillins. The types of *E. coli* which cause urinary infections or septic infection after abdominal surgery are almost as versatile. By far the commonest tests called for in a hospital bacteriology department today are to find which of the available antibiotics a given culture of *Staphylococcus* or *E. coli* is susceptible to. Fortunately, it is still very unusual to find a strain of either organism that is resistant to all the commercially available drugs.

One of the most interesting aspects of bacterial resistance was first observed in Japan in 1959 and has since been thoroughly investigated. As might be expected, resistant strains of dysentery bacilli were frequently encountered in hospitals where many cases of dysentery were being treated with antibiotics. What was unexpected was to find the sudden appearance of bacteria simultaneously resistant to *all* the drugs commonly used at that time (sulphonamides, streptomycin, chloramphenicol and tetracyclines) and that *E. coli* isolated from the same patients showed the same resistance. Further investigation showed that some of these resistant bacteria could transfer their resistance to normal bacteria by the process of conjugation that was mentioned in Chapter 3. If we oversimplify as usual, the genes conferring resistance to these four antibiotics are carried in a DNA molecule quite distinct from the larger DNA molecule that comprises the bacterial 'chromosome'. This smaller genetic unit, known as an 'episome', has some resemblance to a temperate bacteriophage in that it has a capacity to attach itself to one or more of the bacterial genes. During conjugation or transduction the episome is transferred, together with whatever bacterial genes it has picked up, to another bacterium. In most instances a random gene carried from one bacterium to another is of no significance, but if one of the genes determines resistance to an antibiotic it may have great survival value. The episome therefore tends to 'collect' those genes which will protect it against the antibiotics that it will encounter in the hospital environment. It is an illuminating example of how what must be an extremely rare occurrence, viz. incorporation of any particular resistance gene into the episome, can nevertheless have such great survival value

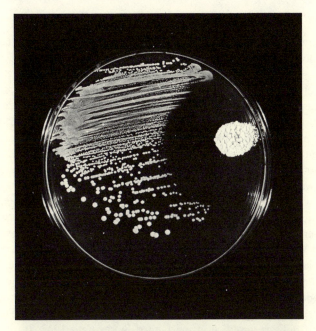

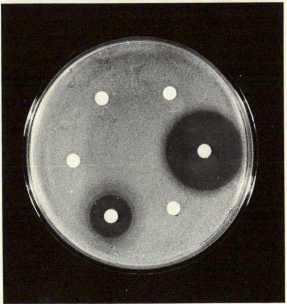

Fig. 21. (*a*) Re-enactment of Fleming's discovery of penicillin: the growth of bacteria on agar was inhibited in the vicinity of a mould (*Penicillium*) which happened to contaminate the plate. (*b*) Antibiotic sensitivity test today: the discs containing particular antibiotics are placed on an agar plate that has been seeded with the bacterium isolated from the patient. The drug diffuses into the agar and inhibits the growth of bacteria. Note that this bacterium is totally resistant to four of the six antibiotics tested.

that the whole population of dysentery bacilli soon carries the quadruple resistance.

Irrespective of the method by which it develops, drug resistance in bacteria is always a potential impediment to any attempts at mass protection against a bacterial infection by chemoprophylaxis. Experience in military camps in 1942–6 provided several cautionary tales which need hardly be retold. Their moral was clear; potent antibacterial drugs must not be used unnecessarily and particularly not in mass applications which will almost certainly facilitate the appearance of resistant strains. There was one situation where it seemed that large-scale chemoprophylactic measures were safe and effective. Meningococcal meningitis has been a notorious plague of army recruits and experience in the Second World War showed that an incipient epidemic could be cut short by dosing the whole military community with a sulphonamide. When everyone was given a daily dose of sulphadiazine for two or three weeks, the proportion of soldiers carrying the meningococcus in their throats fell to a very low level and cases of meningitis ceased to appear. This was standard military practice in the United States until 1962. In that year the meningococci were found to be sulphonamide-resistant and chemoprophylaxis failed. Since that time the principle of mass protection by antibiotics has been abandoned and the rule nowadays is that chemoprophylaxis is only admissible for the protection of an individual known to be highly exposed to some specific infection over a short period of time.

One of the best examples of the legitimate use of the protective power of antibiotics has been their use in the surgery of lung tuberculosis and in any other type of major surgery involving infected or potentially infected tissues. Until the operation became virtually unknown as the result of increasingly successful chemotherapy it was common practice to remove a diseased lobe of the lung under protective cover of streptomycin and penicillin injections. Technically such operations were possible in the days before antibiotics but the mortality was prohibitive as a result of tuberculous or septic infection of the region involved in the operation. The same approach is applied today using the most suitable antibiotics in such procedures as removal of a bowel cancer or open heart surgery.

Another important practical application of the same principle is seen in the penicillin injections which are, or should be, always given when a person with rheumatic heart disease is having teeth extracted. A heart valve damaged by rheumatism is liable to be infected by an otherwise harmless streptococcus that tends to lurk around the apex of an unsatisfactory tooth. Extraction will nearly always result in a little flush of streptococci into the blood, harmless in a healthy person, but if some of them lodge on a chronically injured heart valve a fatal disease, subacute bacterial endocarditis, may be initiated. The risk can be obviated by an injection of penicillin on the day of the dental operation.

*

The antibacterial drugs which have been developed from the two starting points, sulphanilamide and penicillin, have saved many more millions of lives than any other discovery in medicine – but the story has its darker aspects as well. Some of the antibiotics have an intrinsic toxicity which must be balanced against their capacity to cure otherwise lethal disease. Chloramphenicol, with its capacity to deal effectively with typhoid fever and all the rickettsial infections including scrub typhus – most of the infections in fact where penicillin failed – was hailed in 1947 as the perfect complement to penicillin. It was widely used for all sorts of infections and then rather suddenly it was recognized that it caused a fatal anaemia in some children and adults. Those deaths represented only a very small proportion of the patients treated with chloramphenicol but it was enough to call for a strict limitation of the drug to dangerous conditions like typhus and typhoid where it had special virtues. Streptomycin has been a saviour to sufferers from tuberculosis but prolonged administration may cause deafness. Amphotericin B is a highly toxic drug but offers the only hope to patients afflicted with any of the rare fatal diseases due to fungi.

In contrast to these relatively toxic antibiotics, penicillin, with its highly specific attack on the bacterial cell wall, has virtually no intrinsic capacity to damage human cells and is quite remarkably non-toxic in the ordinary sense. Unfortunately however it is very prone to give rise to another type of problem. It may provoke an immune response 'sensitizing' the patient so that when he is subsequently given penicillin for the treatment of some other infection he develops symptoms of hypersensitivity (allergy) which can be quite severe and sometimes fatal.

Penicillin allergy has become very common (5 per cent of people in the USA are said to be sensitive) and unfortunately allergy to any one type of penicillin implies allergy to all the rest. The incidence is increasing, not only as a result of the continuing use of penicillin medically but also because people are being unconsciously exposed to the drug in a variety of other ways. For example, penicillin is used to treat streptococcal infection of the udder in cows, and milk from such cows will contain small amounts of penicillin. Repeated exposure to small doses of the agent is a well known method of inducing hypersensitivity in the laboratory. Of those people who are allergic to penicillin, most discover it the hard way by a sharp reaction to a therapeutic dose. Anyone who does will find that it is wise to carry a card to remind any doctor who may have to treat him in the future that for this patient penicillin is poison.

*

In many ways the administration of an antibiotic to cure an acute infection is one of the simplest of medical procedures. A native orderly with a few months training, in the 'Haus-sik' (dispensary) of some remote New Guinea village, has his anti-malarial pills, his sulpha pills and his penicillin syringes and can use them effectively. We can in such circumstances take many small risks in the certainty of greater overall benefit. In more advanced medical environments we must move more circumspectly, particularly when a new drug is to be introduced – one that is claimed to cure a virus disease perhaps. To test such a drug we must make use of the *controlled trial*.

The controlled therapeutic trial is perhaps the most practically important aspect of modern medical science. In principle the approach is simple. Experiments in animals or some accidental observation in human patients, suggest that some drug or other treatment would be of value to patients with a certain disease. To test it we collect a group of patients, treat half of them with the new drug, half with whatever is the best current treatment of the disease. We tabulate how many die in each group and how long it takes the survivors to get well. The answer should then be obvious.

In fact, things are vastly more complex and very special caution is needed in designing, carrying through and interpreting such trials. No two patients with a given disease are wholly similar unless they are two identical twins. In general they will differ in sex, age, bodily build, past or concurrent illnesses, intelligence and personality. The statistician will insist that the two groups to be compared must be similar in every relevant quality except for the actual procedure under test. In practice this means a large number of patients and allocation to test or control groups by some completely unbiased method.

Where a drug is being tested for its action in controlling some acute infection, a trustworthy answer can usually be obtained without much difficulty. The time taken for the patient's temperature to return to normal is probably the usual criterion. It is objective enough to be trusted and suitable for statistical analysis when the groups are being compared. When penicillin first became available, very few controlled tests were needed to show that it was supremely effective in dealing with streptococcal or pneumococcal infections. The greatest difficulty in such tests was often the ethical one that as soon as a doctor in charge of desperately ill patients is convinced that his new drug is effective, he will naturally wish to give the drug where it is most likely to save life, without regard to whether doing so will ruin his trial or not.

In many clinical trials the result is not a matter of life and death and there is less difficulty in adhering to the planned design. A new difficulty emerges, however, also involving the doctor–patient relationship. It is not always easy for a patient to be sure whether he feels better than he did a week ago, or for a doctor to judge whether his patient is improving or not. If either patient or doctor is enthusiastic for any reason about a new treatment both are likely to be more definite about its benefit than if they were sceptical. To avoid this possibility the 'double-blind' therapeutic trial is now almost standard for any comparison which is expected to show relatively small differences between the new treatment and the old.

As before the available patients are divided into two randomly selected groups. If the drug to be tested is to be given in the form of a white tablet taken three times a day, a large supply of these tablets, sufficient to cover the whole trial, is prepared. An equally large supply of identical-appearing white tablets, which according to the circumstances of the trial are either dummies – placebos, as we say – containing no active drug, or a proper dose of whatever was the standard drug with which the new treatment is to be compared. The two sets of tablets are coded (say A and B) by the dispenser, and which is the new experimental drug is known only to him and a responsible supervisor who takes no active part in the test. This will ensure that neither the patient nor the doctor in charge knows which drug is being used. It is the doctor's business simply to put on record his observations or impressions of each patient at appropriate times throughout the trial period. His ignorance of what is being taken will eliminate any bias for or against the drug. Only when the trial is finished is the code broken and doctor and statistician together can assess the results and their significance. It is all-important that no new vaccine or drug should ever come into use without a trial which is so designed as to give an answer that is unequivocally acceptable to any competent medical statistician and to any physician experienced in the disease being treated.

Trials are concerned only with short-term results. Sometimes the long-term effects may be quite different. The first trials of isoniazid in tuberculosis gave dramatically good results, but the rapid appearance of resistant organisms required a new assessment within less than a year.

Nature can make basically similar responses in other fields. When chemotherapeutic agents are used against malaria or other protozoa, or synthetic insecticides against medically important insects, preliminary success is always liable to be followed by resistance, first patchy then

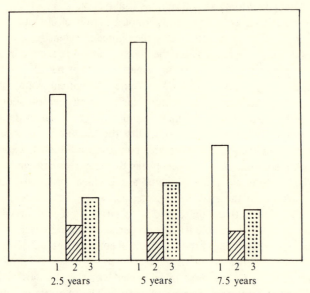

Fig. 22. An example of a controlled trial: about 42,000 adolescents tested for sensitivity to tuberculin and subsequently divided into three groups: 1, negative reactors; 2, negative reactors then inoculated with BCG; 3, positive reactors (these can be regarded as having been immunized by natural infection). The incidence of clinical tuberculosis subsequently contracted in each group is shown at $2\frac{1}{2}$-year intervals. (Based on Medical Research Council Second Report 1959.)

becoming more and more general. Soon a new ecological situation establishes itself and perpetual effort is necessary to maintain the balance at the humanly desirable point.

Another chapter in this story may well be in the making. In 1971 we can report some very limited successes in the chemotherapy of virus disease and a promising approach to a method of stimulating the body to produce the antiviral substance, 'interferon', which appears to play an important part in the body's normal recovery from viral infections.

Neither of these approaches has yet gone far enough to justify any elaboration but we can be certain that even if they do provide some initial triumphs, there will soon be difficulties.

We are moving into a new world in which the old natural history of disease is being rapidly distorted, and we must be always alert to look beyond the immediate effect of some new procedure to see what the logical outcome of its large-scale use will be. Antimicrobial drugs, like measures to prevent the spread of infection or immunization procedures, are potent weapons, but to the biologist they are merely new factors introduced into the environment within which the microorganisms of infection must struggle to survive. We must never underestimate the potentialities of our enemies.

13. Hospital infections and iatrogenic disease

In far too many fields ecology has been ousted by technology. Man is no longer content to allow Nature to work out her own system of checks and balances and to establish stable ecosystems of which he is a part. Whenever circumstances produce a result which seems unfavourable from the human angle, short-term action must be taken to reverse it. With hardly an exception we fail to anticipate some longer-term implication of such action and the repercussions are often unhappy ones. Not a few of the advances of medicine in relation to infectious disease have had to qualify success with some long-term difficulty or accidental disaster.

Hospitals developed historically as places where sick people could be brought together so that their diseases could be studied by scholarly physicians and given the best contemporary treatment. The reality proved to be very different and the hospitals became notorious for septic infection and 'hospital fevers'. In the eighteenth and early nineteenth centuries it was almost a sentence of death to be admitted to a hospital in a European city. Semmelweiss, a Viennese physician, was the first to recognize in 1847 how some of the dangers might be reduced. Puerperal fever following childbirth was rampant in the maternity hospital in which he worked and it was strikingly more frequent in the wards used for teaching medical students than in those used for the training of midwives. A number of circumstances suggested to Semmelweiss that the difference resulted from contagion carried by students from the postmortem room to the lying-in ward where an essential part of their training was the physical examination of women in labour. Institution of a rigid routine, first of soap and water, later of chlorinated lime solution, in which hands had to be thoroughly washed before approaching patients, proved very effective in Semmelweiss's experience. It took many years however before Lister succeeded in imbuing surgeons and others in the hospitals with an understanding of the process by which the medical attendant conveyed potentially lethal infection from one patient to any other patient made vulnerable by injury, surgical incision or childbirth.

Antiseptic and aseptic techniques were developed which progressively made surgery safe from infection. Septic infection by such organisms as streptococci, staphylococci and coliforms however have never been wholly banished from surgical wards and even in these days of antibiotics the problem of hospital infections is still a very important one. Even in the best of hospitals it is unusual for the rate of significant infection in surgical patients to be less than 3 per cent and it is often higher.

The 'hospital staphylococcus' is the greatest danger. It is always penicillin-resistant and in general it will also be resistant to all other antibiotics which have been used extensively in the hospital for more than a few months. Other hospital organisms whose resistance to the commonly used antibiotics has provided major problems of wound infection during the 1960s are coliform organisms from the bowel and *Pseudomonas,* which produces a characteristic green colour of the pus. When the patient is specially vulnerable for any reason such organisms may cause fatal illness – pneumonia or septicaemia. More commonly they produce relatively unimportant infections the practical result of which is to delay recovery by a week or two.

Hospital staphylococci can present a similar nuisance or danger to the newborn infant. In some maternity hospitals it is almost routine for newborn babies to contract staphylococcal infection of the eye, the skin or the umbilical stump. The infection is not infrequently passed on to the mother who can develop a painful abscess of the breast. The source of the epidemic can sometimes be traced to a nurse who is innocently carrying the responsible staphylococcus in her nose. Once infection becomes widespread in a nursery, however, large numbers of babies, mothers and hospital staff can become carriers, shedding staphylococci wherever they go. Eventual eradication rests on such measures as antibiotic treatment of 'shedders', washing of nurses' hands and babies' skin with antiseptic soaps, minimizing the handling of babies by nurses and well-meaning visitors, and so on. But all things considered, the safest place for a newborn baby is at home, and there is an increasing tendency to discharge mother and child from hospital as quickly as possible.

Even more of a problem are the chronic invalids who require prolonged or permanent hospitalization. Now that car accidents have become so alarmingly frequent an increasing number of our hospital beds are occupied by young people with permanent paraplegia, i.e. paralysis of the legs, and of bladder and bowel function, as a result of permanent damage to the spinal cord. Such patients are prone to repeated urinary infections with drug-resistant coliform bacilli which are often almost impossible to eliminate.

So far we have spoken about conditions which are essentially the inevitable result of the organization of treatment within the standard hospital framework, as modified by the use of antibiotics and the establishment of hospital strains of antibiotic-resistant bacteria. Other mishaps may occur in the hospital environment.

The most important is serum hepatitis, a disease that sometimes follows blood transfusion or injection of any material which deliberately or by accident contains human serum. This is of special current interest owing to the recent discovery of a laboratory method of demonstrating the infection. Its detailed discussion is left to a later chapter. Two other infections may be transmitted by blood transfusion but both are much rarer than serum hepatitis. Malaria can sometimes result even when blood from a donor who has been free from malarial symptoms for years is used. Modern open heart surgery demands large amounts of blood, usually obtained from several donors. A rather common pattern of subsequent illness has recently been recognized in patients convalescing after such operations. It takes the form of fever and an increase in the white cells of the blood coming on about a month after the operation. Most of these episodes probably represent infection with the cyto-megalovirus, which has a structural relation to herpes simplex virus and like it usually produces an unnoticed infection in children which can persist indefinitely. Evidence of its presence in symptomless carriers is most easily found in the salivary glands but the transmission of infection by transfusion must indicate that there is a constant leak of virus into the blood. Some of these post-transfusion reactions may be true glandular fever (infectious mononucleosis) the cause of which is not yet unequivocally established though yet another herpes-like virus labelled 'EB' is the current favourite for the role. Only people in apparent good health become blood donors so it follows that accidental transmission of disease by transfusion will almost always be of an agent capable of producing long-lasting symptomless infections.

The next group of iatrogenic infections includes those extremely rare episodes which tend to remain notorious for years, when some vaccine given in good faith proves to contain living virulent microorganisms. We have mentioned in another connection the Bundaberg disaster in Australia where staphylococci contaminated diphtheria vaccine. Other well-known examples will be recalled from earlier sections of this book. They are (1) the Lubeck deaths (1930) in children who were accidentally inoculated with virulent tubercle bacilli instead of the standard immunizing strain BCG, (2) the episode in 1955 when one of the early commercial batches of Salk vaccine against poliomyelitis gave rise to sixty-seven cases of paralysis, which were shown to be due to incomplete inactivation of the poliovirus used in the vaccine; and (3) the 28,000 cases of serum hepatitis with sixty-two deaths amongst US servicemen immunized in early 1942 against yellow fever with a vaccine containing supposedly normal human serum.

In none of these incidents could any blame attach to the persons administering the vaccine; in one way or another the fault lay in the preparation of the material. All of course provoked a strong public reaction and much soul-searching at the technical level. The lessons were well learnt and no episode remotely comparable has been reported since 1955.

There are however dangers arising from faulty techniques of injection particularly when, as often happens, large numbers of people are being injected with the same material. For years it was regarded as an adequate precaution when injecting vaccine or drugs into a group of people to change to a newly sterilized needle for each individual but to use the same syringe for a dozen or more subjects. The commonest mishap was the transfer of serum hepatitis but in one recent incident in Australia two men inoculated with influenza vaccine from a common syringe both died of fulminating streptococcal infection. The danger has been traced to the fact that immediately after an injection has been made and the pressure on the plunger released, there is a small rebound from the tissue where the injection was deposited which can drive a very small amount of the inoculum mixed with tissue fluid and any microorganisms it may contain back into the needle. Either at once or when the needle is being changed some may reach the body of the syringe. In this way the virus of serum hepatitis or any bacteria present on the skin or in the associated tissues of one individual may be injected into the tissues of the next in line.

The accepted routine today is to use a sterile syringe and needle for every individual, and in our affluent society it is now common practice to use plastic disposable syringes and discard them immediately. An alternative where very large numbers are to be inoculated is to use a high-pressure 'gun' to shoot a fine jet of vaccine through the skin. This is quite painless and seems to have every advantage, but though the method has been available for at least ten years not very much use seems to have been made of it.

A totally different type of iatrogenic disease may arise as an accidental by-product of the drug-treatment of some major illnesses. In Chapter 3 we spoke about the 'normal bacterial flora' of the body, the bacteria in the intestinal tract, in the mouth and throat, or on the skin, and about their close relationship to types that can and do produce frank infection of the tissues in the corresponding region. We can go a little further now and speak of many of these 'commensal' bacteria as opportunists ready to become open invaders if the normal defence is let down. They are responsible for trouble in a number of circumstances. Perhaps the best example to start with is not from the hospital ward but from the laboratory. Penicillin is unquestionably the least toxic of antibiotics. Mice, rats and rabbits can safely be given thousands of times the dose necessary to cure an infection. But if a normal guinea pig is given a moderately large dose of penicillin it will die in four or five days from infection by the coliform bacteria normally present in the bowel. This can be traced to and in part explained by the fact that the penicillin destroys a large proportion of the intestinal bacteria – the 'gram-positive' ones – which normally live in a sort of balanced antagonism with the 'gram-negative' coliforms. The coliforms are quite insusceptible to penicillin and given a free field, they overwhelm the guinea pig.

Something similar can occur in human medicine when prolonged administration of a 'broad spectrum' antibiotic like the tetracyclines is necessary. Rather commonly with young children the normal bacteria of the mouth and throat are eliminated by such treatment which gives an opportunity for the yeast *Candida* to proliferate and produce 'thrush', an inflammatory condition of the mouth and throat. Similar changes in the intestine can produce diarrhoea. These occasional complications of chemotherapy should not be made too much of. Most right themselves as soon as the primary infection for which the drug was given has been overcome and the antibiotic is withdrawn.

There are however some conditions which render patients receiving modern treatment so vulnerable to infection that even specially devised 'ultra-clean' wards are not always effective against the opportunistic invaders. They include the chemotherapy of leukaemia and some other types of malignant disease and the use of immuno-suppressive drugs in surgical transplantation. In view of the risk of infection both groups of patients are almost certain to receive antibiotics as well, often of several types.

In dealing first with the risks associated with the use of immuno-suppressive drugs to prevent rejection of kidney or other transplants, we should stress at once that the figures for survival after kidney transplantation are steadily improving and that about a 70 per cent survival rate for five years can be promised by a first-rate centre. Most of the 30 per cent of deaths take place in the first year. This is high, but not in relation to the 100 per cent mortality of the untreated. There is of course the alternative treatment by chronic dialysis (the 'artificial kidney') but this is not the place for us to discuss the comparative virtues of the two approaches. The immuno-suppressive drugs like 'Imuran' and the anti-lymphocytic serum now being widely used both depress the cell-mediated T immune responses more than those in which antibody is concerned. Bacterial infections can usually be controlled by antibiotics and most of the unwanted effects, all of them rare, are related to the weakness of the T system. Cancer, often of lymphoid tissue, appears significantly more frequently than would be expected by chance, though still in less than 1 per cent of patients. There are also on record fatal cases of virus infection by cytomegalovirus or herpes simplex, and of the common but normally benign protozoal infection by *Toxoplasma*. All told the total of deaths attributable to the suppression of immunity is probably less than 2 per cent of the cases of kidney transplantation.

The second group presents a more gloomy picture. These are patients under treatment for inoperable malignant disease. The two important conditions are childhood leukaemia and chorioncarcinoma, a rare but very lethal form of cancer that follows childbirth. The standard treatment for acute leukaemia is cortisone or one of its derivatives and one of the cytotoxic drugs. The difficulty in treating leukaemia in childhood is that the cell which has gone wrong and is multiplying without restraint is almost identical to the precursor of the standard cells involved in the T immune response. The immediate effect of the drugs on the leukaemia is often gratifying but concomitantly vulnerability to virus infection is greatly increased. Measles and chickenpox can both produce disastrously fatal epidemics in the leukaemia ward of a children's hospital. In such patients measles often takes a quite abnormal form in which there is no skin rash and the child dies from 'giant cell' pneumonia. Opportunistic invaders can appear in many guises; when death from blood infection (septicaemia) occurs coliform organisms from the blood are usually responsible, while yeasts and fungi can sometimes produce chronic infection, often after a bacterial infection has been cleaned up by antibiotics. Similar things are likely to happen in such conditions as Hodgkin's disease or chronic leukaemias where the disease itself diminishes the capacity to deal with infection. An American study of post-mortem findings of patients dying with such diseases showed a sharp increase in fungal infections after 1947 when for the first time antibiotic therapy became standard hospital practice for infections.

For most types of cancer, surgery is still the standard primary approach with X-rays to back it up. Only if these fail does chemotherapy normally come into the picture. Most of the drugs used are cytotoxic, i.e. liable to damage any type of cell that is actively dividing. This is also basically true for irradiation with X-rays, and many of the anti-cancer drugs act so similarly to X-rays that they are spoken of as radiomimetic. All anti-cancer drugs are treacherous weapons. A common feature of cancers which makes them susceptible to drug therapy is a relatively active rate of growth, but in bone marrow, thymus and the intestinal lining there are normal cells which proliferate at a speed greater than that of many cancer cells. So in addition to destroying a proportion of cancer cells all anti-cancer drugs are liable to damp down immune responses, to paralyse the bone marrow and sometimes to cause intestinal ulceration and diarrhoea.

The treatment of cancer after surgical eradication has failed will always be difficult and often unrewarding. The task of adjusting anti-cancer drugs to do the least amount of unwanted damage, antibiotics to control the inevitable infections and sedatives to deal with symptoms can prove a nightmare for both physician and patient. Neither of us has responsibility for treating patients – so perhaps we are not justified in asking whether to hold off death for a month or two at such a cost is really offering anything that the patient could desire.

14. Diphtheria

The principles we have tried to enunciate throughout the greater part of this book may be more meaningful to the reader if we conclude with detailed descriptions of the natural history of a selection of the important infectious diseases of man. It is appropriate that we should begin with diphtheria. Other diseases are more important causes of death and some have been just as carefully and extensively studied, but no other common disease has been so successfully studied. From research on diphtheria there was developed a satisfactory method of treating the disease and an outstandingly effective method of preventing its occurrence. In the period 1920–40 while methods of immunization were being devised and tested, much was learnt about the interaction between man as host and the diphtheria bacillus as parasite. It is probably true to say that the detailed studies of the epidemiology of diphtheria during that period did more than anything else to provide a scientific basis for the understanding of infectious disease. It was a model later to be applied to polio, influenza and in fact to every other infectious disease of human significance.

The history of diphtheria is of great interest. Until we reach the beginning of the nineteenth century it is practically impossible to diagnose the nature of the various throat infections which are mentioned in medical writings. The great majority of them were probably of streptococcal nature, including typical scarlet fever and all the various forms of tonsillitis and quinsy. Children certainly died of 'croup', a general term for any form of obstruction to breathing. Most croup nowadays is due to viruses of the parainfluenza group but when diphtheria was prevalent the commonest cause of fatal croup was laryngeal diphtheria. Taken along with other evidence it is reasonable to assume that diphtheria bacilli were common enough but produced only insignificant damage except when the infection spread into the larynx and produced fatal croup.

Early last century definite epidemics of what we can recognize as diphtheria occurred on the Continent, especially in France, Norway and Denmark. The disease was recognized as an infectious one and given its present name by Bretonneau in 1826, but no more than a few stray cases were recognized in England for thirty years. In 1858 there was a sudden widespread appearance of severe diphtheria in England, and within the year it had spread to almost every part of the globe. Even Australia was infected before the end of 1858. The first Victorian cases were in October 1858, and Tasmania experienced the epidemic in January 1859. The small and isolated settlement in Western Australia remained free from diphtheria until 1864, when numerous cases occurred.

Wherever it appeared at this time, diphtheria was recognized as something outside the previous experience of physicians. The spreading grey membrane on the throat, the high fatality and the common appearance of paralysis of the muscles of the palate some time after infection were all new. Nevertheless diphtheria in 1858 behaved epidemiologically like an infection that had long been present in the communities of the Western world. From the beginning it was a disease of childhood, not of adult life. The Australian statistics show that diphtheria was more exclusively a disease of young children in the 1860s than in the 1920s. Even before diphtheria appeared in its classical form children must have been developing immunity against the responsible microorganisms and in 1858 those over ten years of age were nearly all possessed of sufficient immunity to avoid infection. Since 1858 typical diphtheria has been present in all the civilized communities of temperate climates. Its incidence and severity have shown the inevitable ups and downs. There was a second period of high mortality in Europe around 1890, then a steady fall for about thirty years with a distinct recrudescence of activity, in Europe at least, in the period 1927–31. A very sharp increase in the incidence and severity of diphtheria occurred in Germany and the European countries under German occupation during 1941–5, Norway and the Netherlands being particularly affected. This contrasted sharply with the steady fall in diphtheria in Britain during the same period that was associated with, and probably resulted from, a strenuous effort to immunize all children in the country. Since the end of the 1939–45 war immunization has been almost universal practice and incidence has steadily fallen until in most Western nations the disease has now almost vanished.

As soon as the foundations of bacteriology had been laid by Pasteur and Koch, attempts were made to isolate bacteria from the throats of patients with diphtheria. In 1882 it was observed under the microscope that there were numerous bacilli of an easily recognized shape in the grey 'false membrane' which forms on the tonsils in diphtheria. Cultures of these bacilli were obtained without much difficulty a year later. It did not take very long to show that the diphtheria bacillus was the immediate cause of diphtheria, but there was still much to be done before any possibility of controlling the disease could be thought of.

The first step was to show that the symptoms and pathological changes that were produced by injecting cultures of diphtheria bacilli into guinea pigs were mainly or wholly due to a soluble toxin. Filtered material from a broth culture of the bacillus produced essentially the same local changes and in sufficient amount killed the animal. In its time (1884) this was an exciting discovery and it was soon followed by the even more important demonstration that animals could be immunized to produce an antitoxin that would neutralize all the damaging effects of toxin. It seemed clear almost at once that the disease in man must also be essentially poisoning by diphtheria toxin and that it should be curable by the administration of antitoxin, i.e. serum from immunized horses. With some qualifications antitoxin did prove to be effective in countering the disease if given early enough and all subsequent work substantiated the importance of toxin and antitoxin in shaping the epidemiology of the disease.

When the problem of how diphtheria could be controlled was being studied from 1920 onwards, it was necessary to determine the natural history of diphtheria bacilli within the human community, what proportion of children carried the bacteria without symptoms and to what extent children of different ages had already developed immunity. That stage was completed many years ago in most countries but essentially similar information may still be needed when an unexpected outbreak of diphtheria occurs in some unimmunized group of children. 'Swabbing' and 'Schick-testing', two terms not now so familiar to newspaper readers as they were forty years ago, are the two methods mainly used to obtain such information. Swabbing merely means the taking of a sample of mucus from the back of the throat on a piece of sterile cotton wool. The material is then suitably cultured to see whether virulent diphtheria bacilli are present. When diphtheria was prevalent in a city in the days before immunization it was usual to find 2–5 per cent of apparently healthy children with bacilli in their throats at any one time. Since, on average, each individual could be demonstrated to carry the organism for no more than a few weeks it can be calculated that most of them must have been reinfected on numerous occasions throughout childhood. Yet even in those days not more than 5–10 per cent of children ever suffered from clinical diphtheria, so that we can feel sure that on most occasions the presence of diphtheria bacilli in the throat did not produce the disease. If we are to understand the results of these repeated harmless contacts with the bacterium, we must consider what is found by the use of the Schick test.

The Schick test is merely the simplest method of finding out whether or not there is diphtheria antitoxin in the blood. A very weak solution of toxin is injected into the child's skin. If there is sufficient antitoxin in the blood, the toxin is neutralized and produces no result. The Schick test is negative. In the absence of antitoxin, the injected toxin damages the skin cells, and a red patch of superficial inflammation appears after two or three days – a positive Schick test.

Fig. 23 is a graph of the results from a famous investigation made in New York before the widespread use of artificial immunization. This is what could normally be expected from the natural spread of diphtheria bacilli in a crowded city area. It shows the proportion of individuals at different ages who are Schick negative, i.e. whose blood contains antitoxin. Newborn infants are usually Schick negative as a result of antitoxin received from their mothers and it will be seen that nearly 50 per cent still had some of that antitoxin at six months of age. The lowest proportion of negative reactors was reached at about twelve months of age. Thereafter the process of active immunization proceeded as a result of casual, usually non-symptomatic infection by diphtheria bacilli and most children had acquired immunity before they reached their teens.

In all probability diphtheria results only when a child receives either a very large dose of bacilli or a particularly virulent strain before he has had an opportunity to develop an adequate degree of immunity by more harmless contacts. Very wide experience has shown that children who are Schick negative practically never develop diphtheria. They have already sufficient antitoxin to render even a large dose of bacilli incapable of doing any serious damage. It is amongst the children giving a positive Schick test that diphtheria occurs.

It will be easier to understand why there is this difference in susceptibility if we look into what happens when a large dose of virulent diphtheria bacilli lodges on the tonsils of a susceptible Schick positive child who has not previously developed antitoxin. The bacteria multiply first on the surface, but very soon they get into the lining membrane of the throat. Toxin is produced, the first action of which is to kill the adjacent cells and any phagocytes in the vicinity. This provides dead tissue in which the diphtheria bacilli can grow and produce more toxin. More and more damage is done, and soon there is a grey patch of dead tissue over the tonsils and spreading over the palate. There are millions of bacilli here actively producing toxin which now leaks into the blood, causing the general symptoms of diphtheria. The bacilli themselves remain strictly localized to the throat; it is the toxin they liberate which produces general symptoms or death.

196

In a Schick negative child the diphtheria bacilli may establish themselves temporarily in the throat, but the damage they can produce is trivial. The antitoxin in the blood protects the superficial tissues of the throat from toxin so that no dead tissue accumulates to allow free multiplication of the bacilli. With the toxin rendered impotent, local phagocytes eventually dispose of the bacilli in the throat, but it is a process that may take some weeks. Even if with particularly virulent diphtheria bacilli a local foothold is established and the available antitoxin overcome, the partially immune child can very rapidly produce more antitoxin in response to the stimulus of infection, and only a mild attack results.

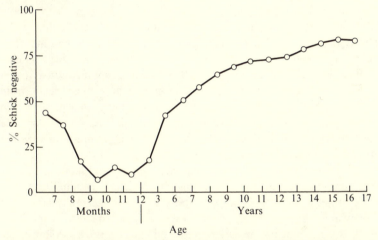

Fig. 23. The development of immunity to diphtheria during childhood, shown as the percentage of children at various ages who are Schick negative, i.e. immune to diphtheria. The immunity of infants is derived passively from their mothers, and disappears during the first year of life. After that time immunity is acquired as a result of infections, which are usually subclinical.

If this is true, it follows that to prevent diphtheria all we have to do is to make certain that each child develops antitoxin before it comes into contact with a large dose of bacilli. We have to find a way of imitating nature's method of harmless contact and immunization, without the liability to serious infection which is inherent in it. The closest imitation would be to inoculate many very tiny doses of toxin but this would be impracticial. Fortunately, it is possible to treat toxin with formalin so that it loses all its poisonous properties but still retains its power to induce the formation of antitoxin. Relatively large amounts of toxoid can be safely injected provided the occasional unduly reactive adult recipient is recognized by a preliminary test. Experience has shown that the standard course of toxoid injections will produce sufficient antitoxin in any normal child to ensure almost complete immunity against diphtheria.

The value of immunization against diphtheria became firmly established soon after its introduction in 1923 and it was gradually adopted throughout the Western world. Fig. 24 shows the trend of mortality from diphtheria in New York City and in England and Wales since 1900. A semi-logarithmic scale is used to give a clearer indication of how great the fall in the death-rate has been. Both curves show first the gradual fall in mortality that resulted from improving living standards and the increasing effectiveness of medical care. Then in New York about 1929–30, and in England and Wales about 1941–2, the line turns sharply downwards. The turning point in both cases is at the time when the proportion of school children who had been immunized against diphtheria reached approximately 50 per cent. Immunization is not a complete protection against diphtheria nor even against death from diphtheria, but the overall effect shown in the figure and paralleled in many other countries must mean that in Britain and North America alone about 30,000 children each year survive who in pre-immunization days would have died from diphtheria.

Today's vaccine is an 'alum-precipitated' toxoid that induces a rather better immune response than the original toxoids. It is administered in two or three injections during the first year of life, generally combined with whooping cough and tetanus vaccines (the so-called 'triple vaccine'). Booster inoculations are required on entering pre-school, primary school and high school.

There have obviously been differences in the virulence of diphtheria bacilli in different periods and in the 1930s much effort was spent in sorting out different 'types' of bacillus according to the type of colony produced, what types of bacterial virus they were sensitive to and so on. When cultures were plated on standard media two distinct types of colony appeared (as well as many intermediate forms) and for a time it was believed that the form of the colonies indicated whether the bacillus in question would produce severe (*gravis*) or mild (*mitis*) infection. There were suggestions too that these differences in 'type' might have been responsible for some of the historical changes in the clinical character of diphtheria such as the world-wide intensification of the disease in 1858–60. If diphtheria had remained an important cause of disease and death such investigations would doubtless have been continued, but in fact they were not and there is no longer serious interest in the once absorbing topic of *gravis* and *mitis* strains.

Laboratory interest in diphtheria bacilli as such has been largely confined to the nature of the toxin and the biology of its production. Its toxic action is directed at the process of protein synthesis in all the cells that it reaches. In technical terms it inactivates the enzyme aminoacyl transferase which has an essential role in stringing the amino acids into protein chains.

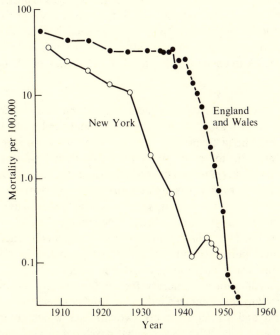

Fig. 24. The death-rate per 100,000 from diphtheria in New York and in England and Wales over the first half of this century. Deaths are shown on a logarithmic scale.

The origin of the toxin-producing capacity was discovered in 1951 in a completely unexpected direction. After a period of incredulity it is now fully accepted that diphtheria bacilli only produce toxin when they are latently infected (lysogenized) by a bacterial virus, diphtheria phage β. This now appears to be completely established although its full signifi-cance must wait on further work by the molecular biologists. Since the diphtheria bacillus is defined as a *Corynebacterium* that produces diphtheria toxin, and we know that there are many different non-toxin-producing 'diphtheroids', it is a reasonable hypothesis that any one of this group of interrelated relatively harmless organisms found as 'commensals' in the human throat will become a diphtheria bacillus when lysogenized by phage β. It is entirely possible that the toxin is genetically determined by the virus rather than by the bacterium and even that toxin may be produced only by those particular bacilli in the culture tube or in a child's throat that are dissolved (lysed) by active β phage. Instead of a simple host–parasite relationship between man and bacterium we have therefore two interacting sets of host–parasite relationships of fascinating complexity. It is ironic that as far as we can see at the present time the production of toxin by the lysogenized bacillus has no discernible relevance to the survival of either virus or bacillus. There was an almost Shakespearean irrelevance in all those deaths of children from diphtheria.

Despite the new discoveries the practical approach to diphtheria is just what it was in the 1930s. Toxin (even if we should call it phage β toxin instead of diphtheria toxin) still dominates the whole picture. All the symptoms are due to toxin, and the patient should be treated with antitoxin – the earlier it is administered, the better the results. Delay is as dangerous now as ever it was, and because delay in treating diphtheria has always been hard to avoid, there has been little change in case fatality in the past fifty years; 5–10 per cent still die. Toxoid immunization in childhood is still routine preventive medical practice. The resulting development of antitoxin not only protects a child from diphtheria, but also greatly reduces the numbers of bacilli in the throat if diphtheria bacilli gain a temporary footing, so diminishing the child's infectivity for any unimmunized individual he may meet.

Diphtheria and scarlet fever are two diseases which resemble one another in being caused by bacteria that develop a powerful toxin only when they are lysogenized by an appropriate virus. In other respects the bacilli that produce diphtheria are relatively heterogeneous and there are several types of streptococci which can be responsible for scarlet fever. In each case however only the toxin-producing capacity is relevant to the clinical result. The history of the two diseases is what would be expected from this character. For at least two centuries both *Streptococcus pyogenes* and *Corynebacterium diphtheriae* persisted as very common endemic infections of the human throat, producing repeated subclinical infections in childhood with resultant immunity to the effects of the toxin. As a consequence, when striking fluctuations occurred in the virulence of the organisms, presumably as a result of changes in the state of lysogenization by toxin-conferring bacteriophages, the clinically 'new' diseases did not behave like new diseases at all.

15. Influenza

Amongst the important infectious diseases, influenza holds a rather special interest for the average man. It is the one disease of which he is almost certain to have had personal experience during adult life. And nine times out of ten it has been a mild, almost pleasurable experience, an opportunity for an unexpected holiday from work. At the end of the First World War, influenza was responsible for what was probably the worst plague in history, yet despite that calamity, its name still remains something of a jest. In previous centuries, influenza presented an even sharper contrast to the other prevailing infections. When smallpox, typhus, typhoid fever and the like were ever present and often fatal, the sudden appearance of influenza which, as a rule, killed only infants and old people, was almost a relief. Influenza was always referred to in a flippant fashion, and Creighton has collected some of the facetious names which have been applied to it in different periods. In the sixteenth century we find it called 'the new acquaintance' or 'the gentle correction'; in the next century 'the new delight' or 'the jolly rant'. In Horace Walpole's letters he speaks of 'these blue plagues'. It was much the same on the Continent. In Germany, one writer referred to it as the 'Galanterie-Krankheit', while the French name still in use, 'la grippe', appears to have had the same meaning as the German – disease *à la mode*. The term 'influenza', which reached England from Italy, with the epidemic of 1782 – the influence – had probably something of the same significance.

For us the chief interest lies in the great epidemic of 1918–19, but to understand how this differed from influenza before and after, it is necessary to say something about the history of the disease. Until 1933, when the virus responsible was first isolated, it was just as easy to diagnose influenza from sixteenth- or seventeenth-century descriptions as it was to say whether a current epidemic was or was not influenza. The disease had to be diagnosed not so much from the symptoms of the patients, but from the characteristics of the epidemic. Other infections might produce the same symptoms of a few days' fever with catarrh and a tendency for pneumonia to follow in old people. But only influenza resulted in a sudden appearance of many such cases with, after a few weeks' prevalence, an almost equally sudden disappearance of the epidemic. Such qualities are not difficult to recognize in old writings. For instance, there is a contemporary description of an epidemic of influenza in Edinburgh in 1562 which included Scotland's most famous queen amongst its victims. It shows clearly how little ordinary influenza has changed through the centuries.

> Immediately upon the Queene's arrivall here she fell acquainted with a new disease that is common in this towne. called here the newe acquayntance. which passed also throughe her whole courte. neither sparinge lordes. ladyes nor damoysells. not so much as ether Frenche or English. It is a plague in their heades that have yt with a great cough that remayneth with some longer. with others shorter tyme as yt findeth apte bodies for the nature of the disease. The Queene kept her bed six days. There was no appearance of danger. nor manie that die of the disease except some olde folkes.

From such lay descriptions as this, as well as from medical writings, it has been possible to compile a fairly complete list of influenza epidemics in Europe since 1500. There are also scattered earlier accounts of the disease, the earliest recognizable reference in English records being to an epidemic in 1170.

The most striking feature of this historical record is the irregular appearance of universal epidemics, or pandemics, involving the whole of Europe, or in more recent years the whole civilized world. In between these pandemics every country has experienced lesser epidemics, often at two-, three- or four-year intervals. There were several earlier pandemics, but the first one to which we need refer is that of 1781-2. This appeared in Asia in the autumn of 1781 and spread through Siberia to Russia, which it reached in December. Finland and Germany were attacked in February, Denmark, Sweden and England in April, while France and Italy, with most of the rest of Europe, felt its effect during the early summer. In London there was a sharp epidemic, limited to the month of June, and probably responsible for 200-300 deaths. This pandemic was not a particularly fatal one, but had one characteristic which is of interest in relation to the pandemic of 1918-19. It spread rapidly, affecting three-fourths of all adults but a much smaller percentage of children, only 2 per cent of the boys at Christ's Hospital being ill. Another comment by a medical writer of the time is significant: 'Children and old people either escaped this influenza entirely or were affected in slight manner.'

We can pass over the fairly severe epidemics of 1803, 1833, and 1837, and come to the pandemic of 1847-8. This, like that of 1782, came from the east, being active in Russia in March 1847 and spreading westward reached England in November. This winter epidemic in England caused a large increase in deaths, particularly amongst the elderly. During the six weeks it lasted there were about 5,000 deaths which could be ascribed to influenza, including not only those attributed directly to influenza, but also the excess of deaths registered as pneumonia and bronchitis during these weeks. Creighton makes the interesting remark in regard to this epidemic that it had the usual effect of an influenza epidemic in lengthening enormously the obituary columns of *The Times*. The elderly rich, protected from the stresses which sent the old folk of poorer classes to their graves in a fairly steady stream, were liable to die in relatively large numbers when something as unavoidable as influenza appeared. It was the one physical danger to the old against which wealth and position could afford no protection.

From 1848 to 1889 there was a rather extraordinary absence of influenza from England. A small number of deaths each year were registered as due to influenza, but no epidemics are on record. One can hardly imagine that the virus was absent from England for forty years, and it is probable that the minor outbreaks which must have occurred produced few deaths and passed unrecorded simply because of lack of interest by the doctors concerned. During this period there were four active prevalences of influenza on the Continent which failed to reach England. The pandemic of 1889 showed the usual spread across Europe from Russia, and reached England in the first week of 1890. This first wave was relatively mild, but was followed by three other waves of influenza at short intervals. The four waves had their 'peaks' in January 1890, May 1891, January 1892 and December 1893. The second and third waves showed relatively high mortalities, each being responsible for more than 2,000 deaths in London. As in previous epidemics, except for that of 1782, the deaths were predominantly amongst infants and old people.

The year 1890 initiated a period of activity of influenza in England, which, after four waves of the primary onslaught, remained always present with exacerbations in 1895, 1900 and 1908, until it culminated in the great pandemic of 1918–19. Like the previous pandemic, this one came in waves. The number and time of the waves varied in different countries. In England, there was a relatively mild but almost universal summer epidemic, then a lull until the main wave rose in October, reaching its highest mortality in November and December. Then came a diminution in deaths, but another serious rise occurred in February and March 1919. In England there were approximately 150,000 deaths from influenza in this period. The rest of the world, with the exception of St

Helena, New Guinea and a few other isolated places, suffered at least as heavily. Most countries populated by western Europeans had a death-rate of three to five per 1,000. Non-Europeans, on the whole, showed a much higher mortality. South African natives lost twenty-seven per 1,000; in India there were over five million deaths, with rates per 1,000 varying in different parts of the country from four to sixty. The highest mortality of all was recorded in Samoa, where a full quarter of the native inhabitants died.

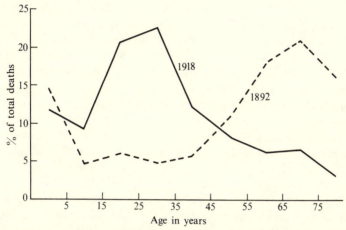

Fig. 25. The change in the age-distribution of deaths from influenza in the 1918–19 pandemic. The percentages of deaths falling in the successive age groups are shown for the 1892 and 1918 epidemics of influenza. In 1892 the chief incidence of death was on old people, in 1918 on young adults.

Such mortalities from influenza were everywhere unprecedented, but an even more remarkable feature was the change in the ages of the people most severely affected. Everywhere the incidence of death fell most heavily on young adults. There were many deaths amongst infants and old people, but the former predominance of these groups in the death-roll had completely vanished. The difference can be strikingly seen by comparing the graph of age-incidence for the 1889–92 pandemic with that of 1918–19 (Fig. 25). This characteristic age-incidence was seen in each of the three waves and in all the countries struck by the pandemic. It had not been observed on any previous occasion, unless we can assume from the brief accounts of the 1782 epidemic that this much less

205

fatal outbreak had a similar age-distribution of deaths. As far as they go, these accounts suggest that the 1782 prevalence was extremely similar to the 1918 summer wave, but was not followed by any severe winter epidemic.

In 1957 there occurred the most widespread pandemic in the history of mankind. At the end of the 1956–7 winter extensive influenza appeared in China and a new type of influenza virus was isolated in Peking. Little notice of this was taken by Western virologists, primarily because mainland China was not a collaborator in the world-wide network of Influenza Surveillance Centres which had been set up at great expense by WHO to monitor outbreaks of influenza in anticipation of just such a pandemic. Several months were lost before the rest of the world became aware of the new virus with the outbreak of influenza in Singapore and Japan in May. All who studied the virus agreed that although it was influenza A it was of completely new antigenic type and that it would certainly spread throughout the world. The pandemic behaved almost precisely according to expectation. Epidemics occurred in the southern hemisphere in June and July and the northern winter showed a double peak of deaths due to influenza and pneumonia in October–November 1957 and February 1958. In both Australia and America there was the expected two years' interval before the next outbreak in May 1959 and January 1960 respectively.

Direct and indirect evidence suggests that between twenty-five and fifty million people died of influenza in 1918–20. Probably a greater number of people were infected with influenza in 1957–8 than in 1918, but the deaths were no higher than those of any 'ordinary' influenza year. There has been much discussion as to the nature of the difference. There are still differences of opinion, but several authorities have been optimistic enough to believe that what made the difference was the advance of medical science in the intervening forty years and in particular the advent of antibiotics. In their view we can feel reasonably confident that, provided our civilization persists, we need not fear that any lethal pandemic of influenza comparable to 1918 will emerge in the future.

There are those who believe that the Roman Empire fell primarily because of inability to deal with outbreaks of infectious disease. With movement of people in war or commerce from one end of the Empire to the other, recurrent epidemics were inevitable in the absence of any of the modern means of handling infectious disease. The development of modern Western civilization went hand in hand with improving knowledge of hygiene. By 1918 the epidemic diseases that killed in classic and medieval times had been brought under control, but not the respiratory infections. The massive troop movements and the upheavals of normal civilian life during wartime provided the best possible opportunity for extensive interchange of air-borne pathogens and there was then no effective counter. There is no substantial reason for believing that the 1918 virus had any greater virulence than that of 1957. Both were 'new' viruses capable of spreading freely through the world's populations. When they killed it was usually as a result of secondary invasion with bacteria. The difference we believe was in the bacteria that were current and in the methods available for dealing with bacterial infection. In 1918 the world was at war, bacterial infections were rife in camps and barracks, in the trenches and in areas of social disorganization. Soldiers with measles died of streptococcal pneumonia, meningitis epidemics were of an intensity never seen in civil life, and there were no drugs effective against bacterial infections. In 1957 the world was reasonably peaceful and reasonably prosperous, and the antibiotics were available everywhere. Mortality was almost limited to older people, particularly those with persistent respiratory ailments, and usually resulted from secondary infection by the notorious 'Golden Staph'. Significantly, there were almost no deaths attributed to the bacterium *Haemophilus influenzae* which had been isolated with such frequency from fatal cases during the 1918 pandemic that it was widely assumed at the time to be the causal agent of influenza itself.

*

This long and puzzling history of pandemic influenza presented the early virologists with a multitude of tantalizing questions about the epidemiology of this strange disease. Their solution had to await the laboratory isolation of the causal agent. By 1933 great advances had been made in the technical side of virus research, and the time was ripe for the discovery of the influenza virus. Many people have heard the story of how it was found by three English scientists that ferrets were susceptible to influenza. They probably do not realize that this was not a chance discovery, but one carefully planned beforehand. If influenza was a virus disease, it could only be investigated by finding an animal susceptible to infection. So when influenza appeared in England in the winter of 1932–3, it was decided to take material from the throats of patients with the disease, filter it through membranes that would hold back bacteria but let any virus pass through, and then test the power of the filtered material to infect all the species of animals that could possibly be used for laboratory work. Rabbits, guinea-pigs and mice refused to show any symptoms at all, so more out-of-the-way animals were tested. The first to be tried was the ferret, because it was known that this animal was highly susceptible to distemper, the virus disease of dogs which has a considerable resemblance to influenza. The ferrets were inoculated into their noses, and two days later they looked miserable, ran high temperatures, developed running noses and sneezed. There was no doubt that they had influenza. That day opened the new epoch in the story of influenza. Now it was susceptible to experimental study, and no longer a nebulous entity that could only be defined by the fact that it occurred in epidemics of a particular type. Since then there has been steady progress in the development of technical methods for handling influenza viruses, so much so that we can say today that influenza A ranks with polio as the most thoroughly studied of all the viruses.

The first technical advance was to transfer the virus to a much more convenient experimental animal than the ferret. The white laboratory mouse is not readily susceptible to infection with influenza, but by transferring the virus rapidly from one mouse to another, mutants can be selected which cause a form of rapidly fatal pneumonia. For several years most experimental work on influenza was carried on by this method but in 1940 the mouse gave way to the chick embryo as the standard 'experimental animal'.

The 'egg techniques' revolutionized the study of influenza and for a period of ten to fifteen years made it technically the simplest of all viruses to study. To grow influenza virus we merely inoculate a minute amount of virus into the amniotic or allantoic cavity of a fertile egg that is halfway through incubation. The virus multiplies in the cells of the surrounding membrane. After two days' incubation we can withdraw the fluid from the cavity and find that it now contains a million times as much virus as we put into it. To measure the amount of virus is equally simple. We merely mix graded amounts with a suspension of red blood cells in salt solution. If virus is present in sufficient amount the cells will be clumped, or agglutinated, and settle to the bottom of the tube in a characteristic pattern (haemagglutination). For virtually every other group of viruses virologists have now adopted cell culture techniques almost exclusively, but for influenza the embryonated egg has remained the most useful means of isolating and cultivating the virus.

Influenza viruses are not identical but differ in respect of several major antigens. Dozens of known strains are classified into two 'types', A and B, based on a difference between the two in the internal antigen that protects the viral RNA. Within each type the several strains are separated by differences in surface antigens, particularly the very same antigens, the 'haemagglutinins', as are responsible for attaching the virus to red blood cells.

Influenza A strains are responsible for more epidemics than B and the outbreaks tend to be larger and more severe. In the northern hemisphere the relatively severe epidemics of 1932-3, 1936-7 and early 1951 were all due to influenza A. Then in 1957 came the pandemic of Asian influenza, again a group A virus, but showing a sharp antigenic difference from all previous strains of this type – so much so that it is known as A2. A variant of A2 (the Hong Kong strain) produced another world-wide epidemic in 1968. No one has any doubts that the 1918-19 pandemic was due to influenza A although direct confirmation is impossible in retrospect.

Viruses identifiable as influenza A have been shown to cause disease in horses, swine, ducks, fowls, turkeys, terns and several other species of birds. It is rather a paradox that away back in 1901 the virus of a fatal disease of poultry, fowl plague, was isolated and has been studied ever since, but not until 1957 was it recognized to be a variety of influenza A virus. Swine influenza as it occurs in Iowa and other mid-western states is of particular interest in relation to the 1918-19 epidemic. It is undoubtedly due to an influenza A virus and farmers in Iowa are confident that the disease they call 'hog flu' was first seen in October 1918 when the great pandemic reached America. They believed that the swine had contracted the human disease and most workers on the subject would now agree that the farmers were probably right. There are doubtless other type A viruses still to be found in domestic or wild animals. So far influenza viruses isolated from animals and birds have not been positively shown to cause infection in man but there are very close antigenic relationships demonstrable in the test tube.

*

With this background we may attempt to interpret what has happened in regard to human influenza during the twentieth century. Even before the days of intercontinental travel by air, human movement had made the whole civilized world a single epidemiological unit as far as influenza was concerned. Wherever there was a large human population susceptible to infection by an influenza virus active in some other part of the world, one could be certain that within a year the virus would produce an epidemic in that population.

A composite picture pieced together from many observations in the past thirty-five years must be based on the changing antigenic character of the influenza A viruses responsible for successive epidemics. Some-

Natural history of infectious disease

times the difference from a previous epidemic was quite slight; at other times, as in 1957, there was a more far-reaching change. Perhaps we can picture what is happening by looking at some large medically advanced country at a time when there are no influenza epidemics. Even in the warm months careful and extensive scrutiny will uncover an occasional case. Influenza A is always present in every major area, but an epidemic arises only when three conditions occur together: first, suitable climatic conditions – influenza is a winter disease as a rule; secondly, large numbers of non-immune people; and thirdly, the presence in the country of a suitable strain of virus. Once an epidemic of influenza A has passed through a district, high levels of IgA will be present in the respiratory mucus of the majority of people and it will be impossible for a similar strain of virus to provoke another epidemic for about three years. If, however, the virus passing irregularly but persistently from one region of the world to another should undergo mutation to a partially new antigenic character it will spread more readily. Two years after the last epidemic the new virus may be capable of overcoming a residual immunity that would have blocked any significant spread by the parent strain. There is a constant premium for survival placed on the emergence of antigenic novelty. So we find a progressive change in antigenicity which we call antigenic drift.

The phenomenon of antigenic drift can be mimicked artificially in the laboratory. If we grow influenza virus in the presence of antibody at a concentration just below that needed to inhibit growth completely, we can select out a mutant which is resistant to that antibody. Furthermore, a close examination of the strains of influenza A isolated from successive epidemics between 1933 and 1956 reveals that each new strain is closely related to its predecessor but differs slightly in one of the surface antigens. The theory of antigenic drift therefore now rests on quite solid evidence and most virologists accept it as the most plausible explanation of these systematic changes.

Such was the simple and satisfying explanation of the origin of human influenza strains up till 1957. Now the whole subject has been cast into a state of flux once more by the discovery that the Asian strain (A2) is totally unrelated to all other recent strains of human influenza – so different that it could not have arisen by a simple mutation. The idea developed therefore that Asian flu may have sprung from some animal reservoir. It had previously been suggested that the 1918 pandemic strain may have originated in the swine of the US mid-West. Now people are asking whether the 1957 strain may have emerged from some reservoir in the Asian steppes rather than by mutation from the A virus current in

210

the preceding human epidemic. It is conceivable that extensive tests for antibody against Asian influenza in the mammals and birds of central Asia could provide a positive answer.

A third possibility is currently being canvassed. Between fifteen and twenty years ago the senior author devoted a great deal of effort to demonstrating that human viruses could undergo genetic recombination

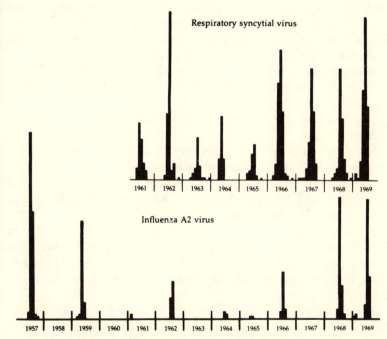

Fig. 26. The incidence of influenza A2 in Melbourne. Isolation of viruses at Fairfield Hospital for Infectious Diseases from 1957 to 1969. The figure shows the primary impact of A2 in 1957 and subsequent recurrences at two- or three-year intervals until the appearance of the Hong Kong strain which was active both in 1968 and 1969. By contrast there have been regular winter epidemics of infection by respiratory syncytial virus in infants. (From Fenner & White, *Medical Virology*.)

and that influenza surpassed all other viruses in its remarkable capacity to do this. The phenomenon can be described in slightly oversimplified fashion as follows. If a cell is infected with a strain of virus that contains the genes A, B, C, all the progeny resemble the parent A, B, C. But if a cell is simultaneously infected with two different strains, say ABC and XYZ, some of the progeny will inherit genes from *both* parents, e.g.

211

ABZ, AXY, etc. Such 'genetic recombinants' are stable and 'breed true' thereafter.

The possibility exists that genetic recombination, so clearly demonstrable in the laboratory, may also occur in nature. It is conceivable for example that somewhere in the world an individual could be exposed simultaneously to two different influenza strains. One might even have been derived from infection in some domestic animal or bird. The resulting recombinant, with properties derived from both parent strains, would represent an entirely new virus, perhaps with pandemic potential. Active work on the implications of this idea is going on in the laboratories. To us it seems a highly dangerous activity only to be carried out under conditions of strict quarantine and perhaps better not done at all. Research on the production of influenza viruses of new pandemic potentiality may be considered to fall into the poet's category of 'lust for knowing what should not be known'.

The saga of influenza is far from over. A vaccine is available but it reduces one's chance of catching flu by only about half. This hardly measures up to our expectations for a modern vaccine. Furthermore, the remarkable propensity of influenza to undergo antigenic drift ensures that in any major new pandemic the virus will be one step ahead of the vaccine manufacturers. The main function of the WHO Influenza Surveillance Centres today is to isolate and recognize a novel viral mutant promptly enough to enable the pharmaceutical companies to make a new vaccine specially 'tailor-made' to meet the demands of the times.

In the continuing battle for survival the virus of influenza has chanced on a near-perfect mechanism for ensuring the continuity of its species. Man now knows enough of the mechanism of antigenic drift to be able to observe the progress of its evolution almost year by year. Clearly if we are to be any match for this formidable adversary we must remain constantly on the alert for the next move in the game.

16.   Tuberculosis

Most of our discussion of the defence processes against infectious diseases has been concerned with the rapidly acting acute infections in which, as a rule, there is a short severe conflict resulting either in death or complete recovery. Tuberculosis is the great example of the chronic infections, and although the same general principles are involved, their application is greatly modified by the slowness of most of the processes concerned.

The tubercle bacillus is a very inactive organism compared, for instance, with the bacilli of typhoid fever or diphtheria. When a suitable supply of food is provided, ordinary bacteria multiply rapidly; a single organism grows to double its original size and divides into two new bacteria every thirty minutes or thereabouts. The tubercle bacillus takes about a day to go through the process that takes the typhoid bacillus half an hour, and as we might expect, the rate at which it produces symptoms in infected animals or men is correspondingly slow. There is one important characteristic of the tubercle bacillus that is probably responsible for much of its behaviour. Its surface is mainly composed of a peculiar waxy material. This wax sheath largely replaces the polysaccharides and proteins of the ordinary bacterial surface, with three important effects: (1) it slows down the entry of food molecules into the substance of the bacterium, hence the slow multiplication of the bacillus; (2) it protects the bacillus to some extent from digestion when it is taken up by phagocytes in the body; and (3) it is less irritant to the tissues than the surface of rapidly multiplying bacteria, and produces a much more slowly acting inflammatory response, involving different cells from those that deal with acute infections. In addition to this waxy material, there is another substance of great importance in the tubercle bacillus, a protein which was first extracted in an impure form by Koch, the discoverer of the bacillus, and called by him tuberculin. There are of course many other substances in the bacillus, but these two have particular importance in relation to disease.

Someone said that of all the millions of species of living organisms on the earth the two about which most had been written were man himself and – the tubercle bacillus. It is therefore no simple task to condense what is important about tuberculosis into a few pages, and for the most part we shall deal only with those aspects which bear on the natural occurrence of the disease.

There are two types of tubercle bacilli which may cause human disease. One is responsible for almost all cases of tuberculosis of the lungs, and is called the human type. The other is primarily a parasite of cattle, and until about thirty years ago was commonly responsible for the tuberculosis of lymph glands and bones in children. This bovine type of the tubercle bacillus almost always entered the body in milk and infected by way of the throat or intestine. With the progressive eradication of bovine TB by slaughter of infected cattle, and with universal pasteurization of milk, these infections have vanished. Although they produced prolonged illnesses, on the whole they were not highly fatal and patients were not infectious to others. The lung infections with the human type of bacillus are much more important practically, and from our point of view much more interesting. We shall, therefore, say nothing further about the bovine form, and concentrate on the endemic disease, pulmonary tuberculosis.

Let us consider the sequence of events when a child inhales infectious droplets coughed up by a tuberculous contact. In the lung a little spot of mild inflammation develops where the bacilli multiply, and from there some of them are carried to the lymph glands which are situated in the centre of the chest where the main bronchi and blood vessels enter the lung. The lymph glands also become inflamed and enlarged, but as a rule the child shows no symptoms and soon the little damaged patch in the lung begins to heal and the lymph glands to return to their normal size. Sometimes healing is delayed and an X-ray will show a shadow in the lung – the primary complex so-called – but even these mostly vanish under treatment or even without. Only a tiny proportion develop frank tuberculosis. In one careful study in midwestern America of 2,266 young people who were known to be infected, only fourteen developed recognizable disease and there were no deaths. But of course there has always been a proportion of infected people who develop active tuberculosis and something must be said about what happens in this small minority of individuals who do not succeed in throwing off their infection.

First of all, there are those whose primary infection fails to heal. Infants born of tuberculous parents are very likely to be infected in the first months of life and die of a generalized tuberculosis in which the bacilli have passed to many parts of the body, especially the meninges of the brain. In other children smaller numbers of bacilli get into the blood from the primary seat of infection and set up a new infection in one of the bones or joints, usually the spinal column or the hip. These are amongst the most horrible of all diseases because they leave a cruel and permanent legacy of deformity. Most of the hunchbacks of former times were victims of tuberculosis of the spine, while mentally retarded, epileptic children who were 'lucky' enough to survive tuberculous meningitis usually spent the rest of their days as 'vegetables' or in mental institutions.

In a proportion of infected children who have developed a primary complex in the lung but who have made a good recovery, a new period of danger may start at adolescence. It may be that in the scarred relics of the primary spot of infection in the lung there are still some living tubercle bacilli. For some reason these are apt to show fresh activity in early adult life, and the scarred area may break down, liberating rather large numbers of bacilli into adjacent parts of the lung. The results of this spread will depend on many factors, but will certainly be influenced by the fact that the lung cells have been rendered sensitive and reactive to the tuberculin protein by the primary infection. When relatively large numbers of tubercle bacilli find their way into the lung of an individual sensitized by previous infection, irrespective of whether they come from an old area of infection or from outside, the reaction of the lung is much more violent than it would be in an unsensitized person. The blood-vessel walls let out cells and fluids into the tissue, which becomes swollen and airless; the pressure of the cells may close down many of the blood vessels, so leaving the area insufficiently nourished.

If, as most pathologists believe, the sensitization process is in part at least responsible for the damage that the tubercle bacillus can inflict in the lung, we may well ask whether this process of sensitization plays any useful part whatever in dealing with the infection. The question was once a hotly controversial one. Today's answer would probably be that conditions are far too complex to state dogmatically whether sensitization is 'good' or 'bad'. It exists and is important in many different ways. The doctor and the pathologist must understand and take account of its potentialities – but there is no call for either to regard the process of sensitization to tuberculin as 'designed' by Nature to help cure tuberculosis.

It would be outside our field to go any further into the complicated details of the struggle with the bacilli in the lung of a tuberculous individual. If we realize that the tubercle bacillus multiplies only slowly, but on the other hand is difficult to kill and may remain present and alive in scarred lung tissue for years, and that the process of cell sensitization

215

may sometimes prevent and sometimes favour the spread of infection, we can see how readily small factors might sway the balance one way or the other.

Probably the most important of these factors are in some way related to the genetic constitution, the inheritance, of the person concerned. In a now famous paper, Kallmann and Reisner described their investigations on tuberculosis in New York State, using the only really effective method that is available for studies of inheritance in human beings. In this study of tuberculosis in twins, as in all similar studies in medical aspects of human genetics, the method was to sort out from the register of cases of tuberculosis all those patients who had a twin. The families of these 'index cases' were then studied in regard to incidence of tuberculosis, not only the co-twin but father and mother, other children and marriage partners. The twin pairs were closely examined to decide whether they were one-egg (identical) or two-egg (non-identical) twins. There were thirty-nine pairs of one-egg twins in the series and, as is always the case, a much larger number of two-egg twins. We can tabulate the results very simply in terms of the percentage of relatives of the index cases who had clinical tuberculosis:

Relation to index case	% with tuberculosis
One-egg twin	87
Two-egg twin	26
Other brother or sister	26
Marriage partner	7

In addition to the 87 per cent concordance in the fact of clinical infection, identical twins also showed a close resemblance in the type and progress of the disease.

Several other factors have a marked effect on the probability of contracting tuberculosis and on the progress of the infection. There have always been striking correlations with age and sex but interestingly they are rapidly changing. Fig. 27 demonstrates how the decline in overall mortality from tuberculosis over the last century has been accompanied by an equally dramatic change in age and sex incidence. Classically, deaths occurred predominantly in infants under the age of three, and in adolescents and young adults, particularly women. Today most of the cases and the deaths in Western countries occur in old men, often derelicts and drunkards from the city slums. Prospective skin testing surveys have shown that almost all of these patients have been tuberculin positive for most of their lives. In other words they were infected first as children or young adults and have carried the bacterium in their lungs

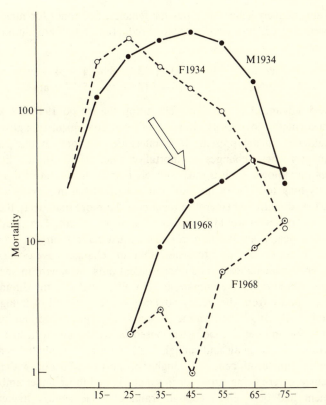

Fig. 27. Changes in the age and sex incidence of tuberculosis between 1934 and 1968. Number of deaths in ten-year age groups in Australia for males and females. Logarithmic scale.

ever since. Reactivation of this 'endogenous' infection has been brought about by the severe stress associated with alcoholism, malnutrition and sometimes associated pulmonary disease, e.g. silicosis.

*

The decade 1950–60 was a turning point in the history of tuberculosis. In every advanced country of the world sanatoria for tuberculosis began to close their doors and public health authorities to feel that full eradication of the disease had become a legitimate objective. This was only the last step in a process that began about a hundred years earlier.

In 1850 the mortality from tuberculosis in England and Wales was about fifty times what it was in 1959. Roughly speaking, the rate has been successively halved six times for females, five times for males, and the periods needed for each successive reduction to half are illuminating. They are:

For females   40 – 30 – 22 – 8 – 4 – 4 – years
For males   55 – 26 – 18 – 5 – 5 – years.

In every advanced country the fall during the period 1946–59 was to unprecedentedly low levels, and the fall has continued, though not so dramatically, to the present. It is interesting to look at the possible reasons for these changes in mortality since 1850. In the light of present-day knowledge it is most unlikely that medical treatment as such had anything to do with the slow but persistent fall in mortality up to 1939. It is doubtful whether treatment ever did more than delay the fatal event in those who would have died without treatment. The steady fall shown in nearly all Western countries must have been due to other factors. There is no reason to believe that any changes have occurred in the tubercle bacillus itself. The improvement must be sought in social or biological factors on the human side. In all probability the diminution resulted mainly from the steady advance in the standard of living over the period. By 1939 the average person in a civilized community was eating more and better food, was housed in greater comfort, had more opportunity for fresh air and sunlight, and was more cleanly in his habits than in the nineteenth century. A higher proportion of people with active tuberculosis were being cared for in sanatoria and those under ambulant treatment had been given enough training in elementary hygiene to diminish their likelihood of infecting others. The net result was probably that on the whole children when they were infected received a smaller dose of bacilli on the average and could deal more effectively with the primary and any subsequent infections.

A good deal of light is thrown on these questions by the facts that have been accumulated about tuberculosis in people who have not previously been exposed to infection. Research on this question became a matter of economic importance with the increasing exploitation of African natives either as soldiers or mine labourers by Europeans. During the war of 1914–18 very large numbers of West African troops and labourers were brought to France, forced to live in a crowded and totally alien environment, and exposed, most of them for the first time, to the tubercle bacillus. The death-rate from tuberculosis was enormously higher than amongst white troops, and the disease was usually far more rapidly fatal. In these non-immune Africans, tuberculosis closely resembled that of those young European children who succumb to their primary infection. The first stage of infection showed little in the way of symptoms, but the bacilli became strongly established in the lung. Sensitization developed rapidly with the production of antibodies against the protein of the bacillus. In Chapter 6 we said something about the increased responsiveness of adult tissues to harmful agents as compared with those of children. These Africans, meeting the tubercle bacillus for the first time in adult life, developed an abnormally high degree of sensitization. If they were sent back to Africa soon after their primary infection, most of them recovered. In France, hard work, exposure and the strange environment eventually resulted in a recrudescence of activity in the lung. The original area of infection set free numerous tubercle bacilli, which were disseminated through the rest of the lung. In their highly sensitized state, the lung cells responded violently, and instead of the chronic changes seen in a European consumptive, these Africans died of what was really an acute tuberculous pneumonia.

In South Africa cases of tuberculosis in Africans employed in the gold mines of the Witwatersrand tended to be of the same general type. Forty years ago the mortality from tuberculosis in mine workers was about five per 1,000 per annum. But for the modern developments in treating tuberculosis, we can be reasonably certain that industrialization in South Africa would have subjected the Africans to the process of 'weeding out' those stocks with an inheritable susceptibility to tuberculosis. Amongst the American Indians this process seems to have been accomplished, and after three or four generations of intense infection and high mortality they now appear to be only a little more susceptible than Europeans living under comparable conditions. Until Mauritius was discovered and settled by the Dutch in 1598 it had no human inhabitants, but Indian and African labourers or slaves soon established an 'indigenous' population. Since then they have been intensely exposed to tuberculosis. Study of their history seems to indicate that it takes something over a hundred years after its first contact with tuberculosis for a race to develop a resistance against the disease equivalent to that of a European population. The process of childhood immunization plays a vital part, but it cannot be fully effective until the more susceptible strains in the population have been eliminated by a very direct form of natural selection.

The precipitate fall in deaths from tuberculosis in the last twenty years is primarily due to the development of effective drug treatment of the disease. Improvements in living standards and in cleanliness continued to play a part and the success of drug therapy had a number of secondary effects that have further improved the situation. In the first place surgery of the lung now became relatively safe and could often greatly speed recovery. With the recognition that tuberculosis could be cured almost with certainty, plus the development of ways of reducing the financial burden of hospitalization, patients became willing and even eager to submit to treatment. One of the results in Australia at least has been that despite the tremendous fall in mortality there are still nearly as many notified cases of infection. In the old days most of these would never have been found.

By the mid-1950s the drug treatment of tuberculosis had become fairly standardized. Three drugs were used, two simple synthetic compounds, para-aminosalicylic acid (PAS) and isoniazid, and the antibiotic streptomycin. Isoniazid and streptomycin when first tested showed dramatically favourable initial results but very soon organisms resistant to the drug being used appeared, often with a relapse in the patient's condition. The difficulty has been overcome by always using more than one of the drugs at any given time. The reason for this is easy to see if we accept the fact that resistance arises by mutation. To simplify the explanation let us assume that whenever we have a population of a million bacilli in a patient's lung one will have mutated to become resistant to streptomycin (S) and one to be resistant to isoniazid (I). If he is given drug S all the bacilli are killed except for the one S-resistant mutant. This multiplies despite the drug and eventually we have a new population of a million S-resistant organisms. Amongst these will be the random mutant that is also I-resistant and if we now change to drug I we may soon find that the patient is infected with doubly resistant bacilli. On the other hand, if we hit with both S and I together there will only be a chance of one in a million million that a spontaneously occurring doubly resistant mutant will be present.

Isoniazid is a non-toxic drug which can be given orally for long periods without close medical supervision and much use of it is now being made as a chemoprophylactic. In careful studies of people who have just become tuberculin positive and are therefore possible candidates for pulmonary tuberculosis, groups given daily doses of isoniazid over some months showed a 50 per cent reduction in the appearance of overt disease. There is some hesitation about applying this method of chemoprophylaxis to children in heavily endemic areas of developing countries, mainly owing to doubt about its effect on the process of natural immunization. More reliance is being placed on the use of vaccination with BCG.

The combination of social progress and effective chemotherapy has

220

made tuberculosis relatively unimportant in Western countries but there is still much to be done in poverty-stricken areas of the world. It will still be many years before tuberculosis becomes unknown in even the most prosperous countries and until then there will be the need to consider protection of those who have to care for or in other ways come into contact with 'shedders' of tubercle bacilli. Prevention of tuberculosis calls for some discussion of tuberculin testing, of case finding by mass radiography and of the use of BCG vaccination.

Whenever information is wanted about the way the tubercle bacillus is spreading in a community, it can be obtained most readily by regular six-monthly or yearly Mantoux tests for sensitization to tuberculin protein. When an individual previously non-reactive shows a positive tuberculin test we know that in the interim he has developed at least a minimal infection with the bacillus. To understand this we must look at some of the things which happen during the primary infection. Tubercle bacilli multiply and are broken down, liberating in the process the protein, tuberculin. As the result of a complex process still incompletely understood the body becomes sensitized to tuberculin. The reaction is in many ways the prototype of what an immunologist calls delayed hypersensitivity and as was discussed in Chapter 6 it is mediated by the lymphocytes we call thymus-dependent immunocytes or T cells. They carry on their surface antibody-like receptors which can react with tuberculin. When this happens they are immobilized and liberate soluble agents which in their turn damage blood capillaries in the neighbourhood.

The Mantoux test is a practical way of detecting whether or not a child or adult has ever undergone this process of sensitization by the tubercle bacillus. For the test, a minute, carefully measured amount of purified tuberculin is injected into the skin. If the person has not been infected, nothing happens. If he has, the tuberculin and the sensitized lymphocytes interact and set going a rapid inflammatory change which shows itself as a red patch in the skin, bright and slightly swollen at twenty-four hours and slowly fading to a light-brownish patch in a few days. It is still routine in some hospitals to carry out such tests at yearly or shorter intervals on all the nursing staff. If a nurse gives no reaction at one test and at the next a typical positive red patch, we can be sure that between the two she has been infected with the tubercle bacillus.

To be Mantoux positive was once considered a good thing. It reflected past infection, hence some degree of acquired immunity to reinfection from others. This protection against the ever-present threat of infection from without far outweighed the risk it carried of a reactivation of disease from within. However, in advanced countries today the risk of encountering *M. tuberculosis* at all is slight. Less than 10 per cent of young adults are tuberculin positive. Any of these are theoretically liable to develop active tuberculosis later in life, but nowadays it is a relatively slight risk if they avoid other causes of ill-health, of which the most important in this context are alcohol and cigarettes.

Prevention of tuberculosis by immunization has had a controversial history and in 1971 vaccination by BCG (bacille Calmette Guérin) in Western countries is virtually limited to persons with no evidence of prior sensitization to tuberculin who are seriously exposed to infection either from contact with a tuberculous patient or relative, or from his profession. The method was introduced in 1922 by the French bacteriologist Calmette as a means of protecting infants born into tuberculous families. The BCG strain of tubercle bacillus is a variant which has lost all power of producing disease on injection into animals, but still has a considerable immunizing power against infection with a virulent culture. Calmette himself gave the vaccine in the form of a living BCG culture by mouth, but today it is always injected into the skin. In several controlled trials there has been a clear-cut advantage on the side of the immunized (see Fig. 22) and in many developing countries there is now widespread immunization of young people with BCG. In many more countries specially vulnerable groups like hospital nurses and medical students are routinely immunized as are also the family contacts of 'open' cases of tuberculosis.

However in areas where tuberculosis rates have fallen to a low figure and where public health authorities are keen to see that every case has early treatment, BCG vaccination tends to be resisted. In such areas it can be claimed that the benefit so far proved is not sufficient to balance the difficulties introduced by the fact that vaccination with BCG provokes a positive tuberculin test. This test has been of great value in the past in assessing the activity of tuberculosis in the community and in detecting cases. Once any considerable proportion of people have an artificially induced positive reaction the test becomes valueless and it is impossible to continue detailed epidemiological work. Even its strongest supporters would probably agree that BCG is only a temporary measure to help accelerate the process of eliminating the tubercle bacillus from the community.

The same can be said of the use of mass radiography in case-finding. If everyone with 'open TB' were segregated from the community till he was non-infectious, tuberculosis would disappear within a decade or two.

Mass radiography is the most effective available weapon for this purpose. The aim is to obtain a miniature X-ray picture of the chest of every adult in the community once a year. Few people have tubercle bacilli in their sputum without some evidence of trouble being visible in the radiograph of their chests. Such screening procedures are now routine and the small number of people with active tuberculosis still

Fig. 28. Tuberculosis survey: a mobile X-ray unit in a remote area of Lesotho. (By courtesy of WHO.)

remaining in the medically advanced countries are usually diagnosed early enough to ensure complete cure by chemotherapy.

Despite the dramatic decline in deaths since the introduction of effective chemotherapy the number of cases notified has not fallen greatly. This is partly accounted for by the recognition of many slight infections and the much greater readiness of patients to present themselves for treatment. Mention has been made of the striking concentration of serious pulmonary tuberculosis nowadays in old men in whom an infection contracted thirty or more years previously is awakened by alcoholism, malnutrition and general debility. In a real sense these elderly male cases are the last indications of the wide, almost universal prevalence of tuberculosis during the days of their childhood. They will probably disappear in their turn in another twenty or thirty years.

*

It is rather characteristic of success in dealing with an important infectious disease that its progressive disappearance brings previously unrecognized infections into prominence. We now know that there are bacteria other than *Mycobacterium tuberculosis* which can produce human infections of the same general character. Some of these 'atypical' mycobacteria come from the soil, some from infections of animals or birds. The human infections are mild – mostly they are subclinical – and infection is not transmitted from man to man.

Their main practical importance derives from the confusion they can introduce into the interpretation of positive tuberculin tests. The 'atypicals' also produce tuberculins which can be used for tests, but they cross-react extensively. Tested with standard tuberculin from *M. tuberculosis*, persons infected with 'atypical' bacilli give reactions which will be read as positive – although strictly speaking they are not tuberculous. There is special interest in the results of some surveys in the US in which highly specific tuberculins were used to differentiate between infections due to 'true' tubercle bacilli and the various atypical strains. Their results showed that in typical American communities only about 5 per cent of young adults these days show evidence of prior infection with *M. tuberculosis* but over one third had been infected with the commonest of the atypical mycobacteria. Of the clinically apparent cases of pulmonary 'tuberculosis' only a minority (perhaps something less than 10 per cent) are due to the atypicals, but as true tuberculosis is progressively eliminated by vigorous chemotherapy, we can anticipate being left with a residuum of infections attributable mainly to this group of organisms.

Perhaps a final word on the possibility of eradication of tuberculosis is in order. Judging from the progress in advanced Western countries this is probably a legitimate objective for the future. It depends mainly on diligent search for cases and vigorous treatment with multiple drug therapy, supported by education, hygiene, chemoprophylaxis with isoniazid, and perhaps BCG vaccination of contacts. The problem is a much more refractory one in the developing countries of Asia and Africa that comprise the main reservoirs of TB in the world today.

17. Plague

For many centuries plague has been the example *par excellence* of the dangerous pestilence. The symptoms are characteristic enough to make it easy to recognize the disease from classical or medieval descriptions, and we can be sure that the two greatest European pestilences, the plague of Justinian's reign (A.D. 542) and the Black Death of 1348, were both the result of the spread of the plague bacillus. Just at the end of the nineteenth century another great plague epidemic arose in Asia, and in India the deaths from plague rose to over a million per annum in 1904. Since then there has been a fairly steady diminution in the annual number of deaths and since 1945 plague has been of little significance as a public health problem in most parts of the world.

From our ecological point of view, it is one of the most interesting of all infectious diseases. Normally, it is not a disease of human beings at all. The great plagues of history were biologically unimportant accidents, the result of human entanglement with a self-contained triangular interaction of rodent, flea and plague bacillus.

Perhaps it would be better to omit the word triangular. Modern opinion is tending to stress the large number of different rodents that may be involved and the different roles they may play in the maintenance and spread of plague. In all probability the steppes and plains of North and Central Asia are the homeland of plague. There it seems to need for its indefinite persistence an association of several rodent species and their fleas. In any area where plague bacilli are to persist indefinitely, there must be a species of relatively high resistance to infection which can serve as a carrier of the bacillus for many months. Alongside the resistant species there must also be a more highly susceptible rodent in which periodic epidemics ('epizootics') will ensure a wide redistribution of the bacillus in the area. The marmots, susliks and tarbagans of Siberia and Mongolia are well known to be subject to recurrent acute outbreaks of plague and as such are dangerous to man. The long-term persistence of infection, however, depends on their association with more resistant but infectible rodents of the gerbille and vole types. Irrespective of the rodent host, infection is spread from animal to animal by fleas. In an infected animal the bacteria are present in the blood and pass into the flea's stomach with its meal. Here, under suitable conditions, the plague bacilli multiply, often sufficiently to interfere seriously with the flea's blood-sucking mechanism. If the flea has in the meantime passed over to an uninfected animal and starts to feed, the blood is not swallowed cleanly, but is regurgitated, mixed with some of the plague bacilli which are obstructing the flea's gullet. The bacilli can then enter the body through the skin puncture made by the flea, and so infect the new host. During the winter, when no detectable cases occur, the bacterium probably survives in the tissues of mildly infected rodents and sometimes in fleas.

We are completely ignorant about the antecedents of the plague of Justinian, but enough is known about the Black Death of 1348 to reconstruct the probable way in which it developed. Probably one of the most important determining factors was the spread to Europe in the thirteenth century of the black rat (*Rattus rattus*). This was originally a native of India, and appears to have reached Europe with returning Crusaders. It rapidly replaced indigenous rats and flourished in and around human habitations. Also, each rat supported a colony of fleas, fleas of a species which had not the exclusive tastes of some such parasites, but were equally ready to feed on rat or man. So conditions were made favourable for the spread of plague when it should arrive. There are several independent sources which locate the origin of the Black Death in southern Russia, and we may be reasonably certain that it was derived ultimately from the natural plague of burrowing rodents somewhere between the Volga and the Don. Probably it was contracted from these rodents by rats and passed to human beings from them. One account says that Italian traders were besieged in a small port on the Crimea for three years until plague broke out amongst their Tartar enemies and caused the siege to be raised, but not before it had been introduced into the town. The Italians returned by ship to Genoa in 1347, carrying the plague with them. Plague broke out in virulent form in Genoa soon after their arrival, and since it is stated that none aboard the ship had shown signs of plague, we can feel fairly confident that it was in rats and their fleas that the infection travelled to Europe. As the Black Death developed its activity, it is almost certain that transmission by flea from rat to man was very largely displaced by man to man transmission, either by way of fleas, or directly when the disease took on the pneumonic form. There is distinct evidence that in the winter 1347–8 the pneumonic form was dominant, while in the summer following nearly all victims contracted the bubonic form.

To understand the differences between the two forms, it will be necessary to make a little digression on the pathology of plague infection.

When a flea introduces the bacilli beneath the skin, most of them are transported to the nearest lymph glands and filtered out there. They are not killed, but multiply, causing inflammation and enlargement of the gland. In a mild case the infection may go no further, but in the severer forms the bacilli enter the blood and pass to other parts of the body. The lymph glands first infected, usually in armpit or groin, however, will in all but the most rapidly fatal cases show swelling and inflammation. There is a hot painful lump, usually discoloured red or purplish. This is the bubo which gives the name bubonic plague to the disease. When the plague bacillus gets into the blood, it will naturally pass to the lungs, and may lodge there, producing pneumonia. In certain circumstances, a patient with plague pneumonia may infect others by droplet infection, the bacteria being transferred directly to the lungs of these new hosts. Here it may produce a primary pneumonia in its turn capable of infecting others. This is the probable way in which an epidemic of pneumonic plague develops. It is the most deadly epidemic disease to which man is subject, but fortunately it appears capable of developing only under certain special conditions.

The Black Death burnt itself out after killing something of the order of one-fourth of the population of England, and for some reason concentrating on adult males. There are figures in regard to the deaths amongst the clergy and amongst the tenants of certain manors which indicate that almost two-thirds of these classes died, but the incidence was certainly much less heavy on women and children. In two parishes mentioned by Creighton, deaths from plague showed sixty-three men to fifteen women in one, fifty-four to fourteen in the other, and there are distinct suggestions that children also escaped lightly.

There was a lull for a few years, but England and Europe were thoroughly contaminated with plague, and fairly frequent outbursts occurred for nearly three centuries. The infection amongst the rats probably never completely died out, and every few years a flare-up of the disease amongst the rodents would be followed by a human epidemic. This went on in England until 1665, the year of the Great Plague. Then after the worst pestilence since the Black Death, the disease disappeared completely from England, until a small outbreak occurred in Essex in 1909. Almost the same thing happened in France in 1720, when Marseilles and the south of France experienced a final disastrous epidemic of plague. It is by no means certain why the long visitation of plague in Europe ended so abruptly, but it is highly probable that one important factor was the replacement of the black rat by the brown or sewer rat. Just as the black rat invaded Europe from India five centuries earlier, so at the beginning of the eighteenth century brown rats began to stream from their original home somewhere in central Asia toward western Europe. Great hordes swam the Volga in 1727, and in 1730 they had established themselves in England. The black rat was very largely displaced from Europe by the newcomer, but continued to be the common ship rat. The brown rat is an objectionable creature, but is not so dangerous a transmitter of plague as the rodent it replaced. It does not frequent dwelling houses so much, and its predominant fleas do not bite man so readily as those of the black rat.

Whatever the cause, plague certainly died out from Europe in the seventeenth and eighteenth centuries, and in the world in general it was very unimportant until 1896. Then an epidemic at Hong Kong marked the beginning of another great phase of plague activity which has not yet altogether ceased. In 1899 it spread to Bombay, and in a few years became one of the most important causes of death in northern and western India. Ship rats and their fleas carried it to every port of consequence in the world, and in many countries it took root, in other words established itself as a parasite of the local rodent population, usually the rats of the waterfront. From these, human outbreaks took their origin.

During the first epidemic at Hong Kong, Kitasato, a Japanese bacteriologist, discovered the plague bacillus. This was the first successful step toward the scientific understanding and sanitary control of plague. During the first ten years of the new pandemic very active research was carried out, and at the end of that period the relation of the rat and its fleas to the disease was well understood and the necessary measures to stamp out, or at least to prevent entry of, plague could be clearly envisaged. It was clearly necessary to attack the rat by every possible method, and above all to prevent the movement of rats from infected to uninfected regions. Measures were devised to clear ships of rats by fumigation, to prevent their movements to and from ships in harbour, and to make grain stores and other warehouses along the waterfronts as nearly rat-proof as possible. Most port authorities kept up a steady campaign of rat destruction with periodic bacteriological surveys to determine whether infection was present in the animals. Such measures have been progressively improved and nowadays much more stress is laid on rat-proofing ships and wharf buildings by proper construction, than on destruction of rats. Most authorities now believe that plague is never again likely to cross the oceans. The steamships that spread plague around the world in 1896–1904 have been replaced by ships in which rats cannot exist. However, a new threat is posed by the recent trend towards 'containerized cargo'. Modern ships carry their cargo in large containers that may leave the wharves by road or rail to be packed on site in factories, granaries or warehouses all around the country. New standards of rat-proofing will need to be carefully policed if the near-impeccable record of recent years is to be maintained. Meanwhile, despite the cessation of intercontinental spread of plague by sea since the pandemic at the turn of the century, plague imported from Asia at that time found the appropriate association of rodents that has allowed it to persist indefinitely since.

The story of plague in South Africa is particularly interesting. Like the rest of the world, its ports were infected from India during the period 1899–1905, and plague became prevalent amongst the rats. Other rodents with a range including both town and country were also infected, and plague spread slowly in these. About 1914 it reached sandy country in which gerbilles (a species of larger burrowing rodents) were numerous. Amongst these the infection spread very rapidly, and by 1935 enormous areas of country were potentially infected. Even within such a short period, plague amongst the gerbilles of South Africa took on the status of a relatively well-balanced persistent infection. When seasons are good, the gerbilles multiply rapidly, but as soon as their numbers reach a certain level the plague bacillus spreads more easily, and an epidemic breaks out amongst the rodents. Their numbers are then rapidly reduced, but it is during the period of widespread disease in the animals that human infections are liable to take place. In South Africa there is a curious little zoological complication to the story at this stage. The gerbille is a creature of the veldt, and avoids human habitations, so the chance of human infections being derived from sick gerbilles is remote. Another rodent, the multimammate mouse, acts as an intermediary. This mouse takes the place of the common domestic mouse in South African houses, building nests in the walls and living on household odds and ends, but on occasion it roves further afield. In the bush it frequently enters gerbille burrows, where it is liable to attach to itself fleas from an animal sick or dead with plague. These are brought back to the house, and hence human infection may occur.

Much the same state of affairs had developed in California, which, like so many other parts of the world, was infected with plague about 1900. The waterside rats as usual formed the first reservoir of the disease, and California has the credit of organizing the first thoroughgoing campaign against rats as a means of checking plague. On the whole, the anti-rat measures were successful, but it was disconcerting to find after some years that plague was widespread amongst the ground squirrels. American epidemiologists are still uncertain whether plague reached the ground squirrels from the imported San Francisco outbreak or whether it had been present unnoticed in the rodents from time immemorial. Whatever its origin, there is no doubt that plague finds as congenial a home amongst these American rodents as it does amongst the marmots and tarbagans of central Asia. There are the usual fluctuations in activity from year to year, but on the whole the area involved spreads progressively. The ground squirrels live mostly remote from human habitations, and relatively few human cases can be traced to them. It is rather alarming, however, to find that small epidemics of the very dangerous pneumonic form have been traced to infection from ground squirrels. For some reason, there seems to be a tendency for plague contracted from burrowing wild rodents to take the pneumonic form, while infection derived from rats is almost always bubonic. In modern times the only great epidemics of pneumonic plague have been in Mongolia and Manchuria, where the tarbagan is the important reservoir.

Today plague remains endemic in certain parts of southern and central Africa, South America, western USA and central and SE Asia, recent epidemics having occurred in Indonesia and Nepal. Major urban epidemics in man are a thing of the past, but the threat of sporadic rural infection remains. For example, plague has become quite prevalent in South Vietnam during the protracted war of the 1960s. American troops are routinely immunized with one of the several vaccines now available, but none of these vaccines is thoroughly satisfactory and all require frequent booster injections.

The development of drugs and poisons selectively active against plague bacilli, fleas and rats since the days of the Second World War, has completely changed the outlook on plague. If patients, even those with pneumonic plague, can be treated early and vigorously with streptomycin or one of the tetracycline antibiotics, nearly all will recover. Contacts may be protected by chemoprophylaxis with tetracyclines or sulphonamides. DDT powder in rat runways and in houses will greatly diminish flea populations, and poisoning campaigns with fluoracetate carried out with proper regard to the subtleties of rat behaviour can be extremely effective. No country with a reasonable standard of living and a competent public health service need any longer fear a serious outbreak of plague.

On the other hand, plague is bound to persist in the wild rodents of the steppes, the veldt and the high plains of North America. Plague may also smoulder amongst domestic rats in parts of Asia, notably India, with occasional small human outbreaks. Rat-proofing of ships and waterfront warehouses will probably be effective in stopping movement overseas from these residual foci. Only the complete disorganization of civilization could bring the plague bacillus back as a major threat to human life.

18. Malaria

Of all the infectious diseases there is no doubt that malaria has caused the greatest harm to the greatest number. All over the tropical and subtropical zones, wherever there were aggregations of people, there malaria flourished. In India it was calculated that in 1930 about a hundred million people were infected with the parasite, and that about two million deaths per annum were directly due to malaria. The influence of the disease extends far beyond its obvious activities as a cause of death and serious illness. It was the great devitalizer of the tropics – much of the backwardness of the Indian peasant has been ascribed to malaria – and it was the main agent of infantile mortality all through history till the end of the Second World War.

With the development of effective methods for mosquito control during the 1940–5 period, it suddenly became possible to envisage the elimination of malaria from the world. By 1960 so much progress had been made (see Fig. 17) that the World Health Organization could proclaim the elimination of malaria as one of its practical objectives. Whether the world's political wisdom will be adequate to handle the problems associated with the resulting tremendous increase in the numbers and potential vigour of the tropical and subtropical countries, is more than doubtful. By the Hippocratic oath and by the traditional morality of every country and every age, the medical scientist and the doctor must save every human life he can. All we can do as human ecologists is to underline the implications of overpopulation and press as hard as we can for a sane and realistic approach to the necessity of birth control in all countries and in all religious and ideological groups.

Once again it seems likely that the incidence of malaria is going to play, as it has in the past, a major part in human history. There is good reason to believe that malaria played a major part in the decline and fall of the Roman Empire, of Greece, and of the ancient civilization and power of Ceylon. If we knew more of the evolution and spread of malaria, we might well find that it played an even more important part in the shaping of human evolution and destiny. Whatever its early history, by the time of European sea-power, malaria was rife over the whole of the continental tropical and subtropical belts. Even thirty years ago the idea of its elimination was unthinkable. Man, mosquito and malarial parasites seemed to have reached a balanced interaction involving millions of square miles of territory and hundreds of millions of human beings. To disturb that equilibrium significantly appeared to be impossible. Today we are confident that a reasonable expenditure of intelligence, effort and money can remove malaria from any part of the globe.

Most readers will have a sufficient general knowledge of the nature of the symptoms of malaria – recurring attacks of severe fever, with intervals of two or three days during which symptoms are absent – to allow us to take them for granted. We must, however, say something about the three closely related species of microorganism which are responsible for about 99 per cent of cases of the disease. They are protozoa with a rather complicated life history, part of which is spent in human beings, part in mosquitoes of one particular genus, *Anopheles*. It is natural to start with the mosquito bite, by which a few tiny parasites rather smaller than most bacteria are inoculated from the insect's salivary glands into the blood. Within an hour these are swept out of the blood, finding lodgement in cells of the liver. Here the parasite finds suitable conditions for multiplication and each organism gives rise to a close-packed mass of tiny descendants that distend the cell eventually to breaking point. About twelve days after the mosquito bite the malarial parasites pass from the disrupted liver cells into the blood and the blood cycle that gives rise to the typical symptoms of malaria is initiated. The liberated parasites are small specks of protoplasm that attach themselves to and enter the substance of the red blood cells. Within the cell each invading protozoon feeds on the cell substance and grows almost as large as the cell. Then the parasite rapidly divides into twenty or more, each with its tiny living nucleus, and when the remnants of the cell break down, these progeny are liberated into the plasma. Then the cycle is repeated, each little descendant parasite, known at this stage of its developmental cycle as a merozoite, enters a fresh red cell and in the space of forty-eight or seventy-two hours, depending on its species, goes through its period of growth and division again. When a large number of merozoites are simultaneously liberated along with poisonous products of their growth, the body's response is the sharp attack of shivering and fever which is the chief symptom of malaria. These attacks continue for a time which varies greatly according to the virulence of the strain of malaria, the general health of the patient, whether he has had previous attacks, what treatment he receives, and so forth. If he does not die, a degree of immunity develops; fixed phagocytes, aided by antibodies in the blood, destroy most of the parasites, but in untreated and often in conventionally treated patients small numbers may remain present in the blood and sometimes in the liver or other internal organs for long periods of time, even for years. As immunity develops, a new phase of the parasite appears in the blood. These are sexually differentiated, and their function is to infect any mosquito of the right species that bites the patient at this stage. In the mosquito's intestine, the male and female forms unite, and from their fusion a new brood of parasites develop. These invade the wall of the intestine and grow into little nodules swarming with tiny needle-shaped organisms. When the nodules break down, these 'sporozoites' pass to the salivary glands of the mosquito, and after a period of from ten to twenty days has elapsed since it became infected, the mosquito becomes capable of passing on its infection to another human being. As a reminder of the morphology of the various human stages of development of the *Plasmodium* the reader is referred back to Fig. 3 in Chapter 4.

An important feature of this cycle is the necessity for each stage to be passed through in the appropriate order. The needle-shaped sporozoite from the mosquito cannot enter the red cell, neither can the liver form nor the merozoite infect the mosquito. Even more important, from the practical point of view, is the fact that the different stages are unequally susceptible to the action of antimalarial drugs. In addition each of the three major species of malarial parasite, *Plasmodium vivax, Pl. falciparum* and *Pl. malariae*, has its own points of difference in detail, but there is nothing to be gained by attempting to describe these nor is it necessary to mention other species which on very rare occasions are found to cause human malaria.

Malaria was a military problem of the utmost importance in the Pacific War of 1941–5, and great advances in its practical control were made during that period. Something must be said about the new methods of control, but for the most part we shall be concerned with the natural history of malaria as it existed, or still continues to exist, in regions where no serious attempt to interfere with its incidence is made. In the great majority of endemic areas the two main clinical forms of malaria, benign tertian due to *Pl. vivax,* and malignant tertian due to *Pl. falciparum,* occur together. There are important differences in symptoms and in their response to drugs, but it is legitimate to deal with all types together when discussing the general problem and refer to the complex of infections simply as malaria.

Let us first take a native population in some heavily infected region – northern Ceylon, Malawi or certain parts of Central America. Conditions are suitable for the breeding of large numbers of *Anopheles* mosquitoes, and there is a fairly concentrated resident human population. Infants are infected at an early age, and throughout their lives will be subjected to repeated reinfections. Very many will die during infancy and childhood, some directly from malaria, others from a combination of other infections and malnutrition with the weakening influence of chronic malaria. Those who survive must have perforce acquired a considerable degree of immunity. As we have already mentioned, immunity to malaria is largely associated with the activity of fixed phagocytes, and one of the results of repeated childhood infections is a great enlargement of the spleen. Just what the reasons are for this splenic enlargement is largely unknown, nor is it clear what secondary results derive from the splenic changes. In practice, the degree to which the children's spleens are enlarged in a native community provides a useful index as to how intense malarial infection is in the region. The adults, provided they are adequately nourished, remain substantially immune – infection by an unusually active strain may provoke definite symptoms, but serious results are likely only when there is some failure of food supply. It is of interest that in heavily malarious African regions, although there is an appalling infantile death-rate, the adult well-nourished natives are very healthy. In India, where pressure of population on the means of subsistence is greater, chronic malarial ill-health is much more frequent in the adults.

There are many parts of the tropics where malaria is by no means as universal as in the regions we have described. The right mosquitoes are present, but the factors which make them dangerous are only intermittently present. A very good example of the way epidemics of malaria may affect a tropical population is provided by the story of the Ceylon epidemic of 1934–5. This is said to be the greatest malaria epidemic since accurate records have been available, causing 80,000 deaths amongst two or three million cases.

Ceylon can be divided into a dry northern area, which is highly malarious and relatively sparsely populated, and a large well-watered south-western area, thickly populated and relatively free from malaria in normal years. In 1934 the south-west monsoon and its associated rains completely failed to appear. Crops failed, rivers ceased to flow and became mere chains of waterholes – the climatic conditions approached those of the northern half of the island. The malaria-carrying mosquito of Ceylon breeds in pools of clear water exposed to sunlight, not in

235

overgrown swamps or ricefields nor in flowing streams. With the partial drought, conditions in the south were ideal for its multiplication in the pools along river courses. There was a plague of mosquitoes, and the population was badly undernourished from failure of crops. The conditions made a serious outbreak inevitable, and in October 1934 malaria spread almost everywhere, striking whole villages at once. It was not abnormally fatal – only 3 per cent of patients with symptoms died, but it affected about half the total population. In the northern part of Ceylon there was no abnormal incidence of malaria – conditions remained as before.

In this example, we have a combination of three main factors which favour malarial prevalence. First, unusually suitable conditions for the multiplication of *Anopheles* mosquitoes, secondly, a human population most of whom were not immune – it had been five years since the last serious prevalence of malaria in the region – and thirdly, a population whose resistance had been weakened by undernourishment.

Although malaria is preeminently a tropical disease, it was once very prevalent in England and certain coastal districts of the Netherlands. The recent stages in the process of eliminating malaria from the Mediterranean have been the result of deliberate action but it is more difficult to understand why malaria had so largely disappeared from Europe before the new methods became available. Climatic and social factors have almost certainly been involved. In Italy there have been centuries in which the country was almost completely free from disease and long periods when malaria was rampant throughout the country. It may be significant that the three ages of Italian political and intellectual importance, the first centuries A.D., the Renaissance and the present time, were all periods of low malarial activity. Such alternations probably depended on changes in the mosquito species or races, and particularly on changes in their habits. One of the most interesting and, before the days of DDT, one of the most important phases of malaria research concerns itself with the detailed ecology of the various species of *Anopheles*. There are many parts of the world where there are *Anopheles* mosquitoes but malaria does not spread, even when cases are introduced from the tropics. Sometimes this is because local farming conditions make it much easier for mosquitoes to feed on cattle than on human beings. Sometimes more subtle factors are involved. In Holland around 1930 for instance it was found that there were two structurally similar races of *A. maculipennis*. Although almost indistinguishable in appearance, they had different habits, and while one race transmitted malaria the other did not. The harmful race spent its larval stage in the brackish water of coastal lagoons, the other preferred fresh water. In winter, the malaria-transmitting race spent its time near the ceilings of warm farmhouses and continued to feed on human blood, incidentally transmitting malaria during the process. The harmless race spent the winter in cold outhouses and the like, and hibernated. As exemplifying the difficulties of research on malaria, we may mention that in Holland, while most infections were transmitted by mosquito bite received in this way during winter, the attack of malaria did not occur until the following summer. This very long 'incubation period' has also been observed after experimental infections in man made during research on the treatment of brain syphilis by malarial infection. Such a long delay is probably a function of both the low environmental temperature and of the relatively low virulence of the malarial strain. Ecological work of this sort in Europe played a major part in the cleaning-up process that has now almost banished malaria from the whole continent.

One of the most impressive examples of anti-malarial work was in Brazil, where the introduction in 1930 and spread over thousands of square miles of a dangerous African mosquito was followed by a catastrophic outbreak of malaria in 1938. By this time effective methods of dealing with the breeding places of the mosquito had been developed, and an intensive two years' campaign, even without the new insecticides, eliminated *A. gambiae* from the whole infested region. Similar success was achieved when this same mosquito spread into Egypt and caused a severe epidemic of malaria in 1942.

The requirements for eradication of malaria can be stated as:

(1) the killing of adult female mosquitoes of vector species, especially those likely to have already become infected. This is the rationale for residual spraying in dwellings and is said to be the most cost effective approach to reducing the intensity of malaria,

(2) eliminating all potential breeding places for the vector species,

(3) the treatment of all infected persons at least to the level which will make them permanently non-infective for Anopheline mosquitoes.

The success of such a programme is dependent on the assumption universally accepted before 1960 that the only vertebrate reservoir of human malaria is man. It is known that the most important source from which mosquitoes derive their infection is the blood of indigenous children living in the malarious region. As these children recover from their early infections they carry large numbers of the infectious sexually differentiated forms of plasmodium in their blood.

It was known for many years that malarial parasites carried by

Fig. 29. (*a*) Malaria control: training a team for residual spraying with DDT.

mosquitoes infected many types of bird and mammal in tropical and subtropical lands. Only in 1960 however was it discovered that monkeys in South-East Asia were carriers of a plasmodium which on injection into human volunteers could provoke a mild but typical malarial infection. Since then there has been close study of primate malaria from this point of view. Both chimpanzees and gorillas can carry the parasite of human quartan malaria, and another Malayan monkey parasite has

238

Fig. 29. (*b*) Malaria control: the field laboratory. (Both plates by courtesy of WHO.)

produced human malaria by natural infection in the jungle. Opinion in 1971 was however that malaria of non-human primates presents no real threat to human health. In the absence of a human reservoir there should be only occasional cases of malaria in jungle workers and further spread could easily be checked.

Malaria prevention by mosquito control will always require a thorough knowledge of the ecology of each important species of Anopheline mosquito. Each has its own type of breeding place and its own feeding habits; and no mosquito species will readily change its pattern of living. Systematic attack on the larval stage must be adapted to local circumstances. Breeding places may be drained or altered in one way or another to make them unsuitable for mosquito larvae of the significant species. The larvae can be suffocated by an oil film on the surface of the water, poisoned by Paris green or eaten by small fish. At the human level we can prevent persons being bitten by malaria-carrying mosquitoes by such means as efficient screening of doors and windows, mosquito repellents (e.g. dimethylphthalate), nets at night and protective clothing by day. We can kill mosquitoes in the houses by leaving a fine residual deposit of DDT on walls and ceilings where mosquitoes rest in the daytime. The widespread use of this particular insecticide has led to

239

the emergence of DDT-resistant mosquitoes in many parts of the world, hence alternative chemicals are now being preferred. Finally, by appropriate dosage with drugs, infected persons can be rendered non-infective for the mosquitoes that bite them.

Malaria can be a problem even to the countries that have wholly eliminated it from their own territories. In 1971 malaria is still the tropical disease most commonly imported into Europe and the USA. The explosive increase in the numbers of tourists and others who travel to tropical areas where malaria is endemic makes two requirements mandatory: (1) chemoprophylaxis with chloroquine or proguanil while exposed to infection and for at least six weeks after leaving the area, and (2) medical surveillance for several months after return to the non-endemic area, because clinical malaria can appear after a long incubation period and the diagnosis is often missed unless the physician is kept aware of the possibility.

Given understanding of what is required, well trained technical personnel of many types, and adequate finance, any problem of preventing or treating malaria can nowadays be dealt with successfully. Any tropical city with reasonable financial resources can free itself from malaria in a relatively short time. It was claimed by WHO that when the decision to work for the world eradication of malaria was taken in 1955, 1100 million people were living under threat of malaria. Ten years later the number of people at risk had been brought down to 600 million. Of course there have been difficulties. Governments hard pressed financially have been liable to stop expensive antimalarial measures before the job was complete, and a change in the timing of the rainy season has thrown carefully worked out plans into chaos. One of the major difficulties in recent years, highlighted by the American Army's experience in Vietnam, has been the emergence of drug-resistant strains of *Pl. falciparum*, the parasite causing malignant tertian malaria. It has been common to find strains resistant not only to the standard antimalarial drug, chloroquine, but also to the other major synthetics. By a curious turn of fate the ancient antimalarial drug, quinine, derived from the bark of the cinchona tree, is again being widely used and the newly developed combination of sulphonamides with trimethoprim is providing a hopeful new approach. It is obvious that malaria will require long-continued and determined attack on all fronts before it is eliminated from its major tropical strongholds. It is even more certain that continuing social disorganization by war and revolution will prevent eradication, and any social breakdown in cleaned-up areas will allow a rapid return of malaria.

It is now too late to question whether WHO was right in aiming at world eradication. The die has been cast and every country now demands that malaria be rooted out from its territory. The impact of diminishing malaria on vital statistics is evident everywhere. Mauritius offers a particularly striking example. Between 1945 and 1948 the two malarial mosquitoes were eliminated; birth-rates increased sharply from about 35 to 45, infant mortality fell from 150 to 80, and general mortality from 26 to 12 per 1000. Population, from around 420,000 in 1945, had soared to 750,000 in 1965. The birth-rate had returned to 35 but the death-rate had fallen further to 9 per 1000, giving a population increase still close to 3 per cent per annum. Being an island of less than 800 square miles Mauritius clearly faces a desperate problem of over-population which it has recently begun to tackle. Birth control clinics are now active and in 1971 the rate of population increase was down to 1·9 per cent per annum. It is the same problem that confronts the rest of the world wherever birth control fails to catch up with the surge of fertility and health that has been released by the removal of infectious disease – and particularly of malaria.

19.   Yellow fever

As Greenwood has pointed out, the name of each important disease has its own emotional colour, not necessarily very closely equivalent to the real importance of the disease. One has only to run over in one's mind the names 'plague', 'leprosy' and 'influenza' to realize this. Perhaps of them all yellow fever is the most vividly coloured by associations, quite apart from the image conjured up by the name itself of an intensely jaundiced patient dying in delirium. There has been something grimly romantic about its story from its first appearance, when it seemed to rise like a miasma from the overcrowded stinking holds of the slave ships in Barbados in 1647.

For two and a half centuries, yellow fever held possession of the Spanish Main, slaying English, Spanish and Carib, soldier, buccaneer and merchant alike, but strangely sparing the negro slaves. In time there grew up a local population in the West Indies which seemed to be immune to its attacks but almost every newcomer had to face its peril. At times yellow fever spread widely over all subtropical North and South America, on several occasions causing severe epidemics in the southern United States. On the other side of the Atlantic was the west coast of Africa, the home alike of yellow fever and of the slaves with whom it had travelled to America. Here also the disease, hardly noticed by the native inhabitants, played havoc amongst Europeans and, with malaria and dysentery, was responsible for the evil reputation of the country as the 'white man's grave'. From the west coast it spread northward at irregular intervals, reaching the Mediterranean countries more than once. Portugal suffered particularly severely about 150 years ago.

Something of the same grim romance hangs round the story of the beginnings of modern research on yellow fever in 1900–1. At that time, just after the Spanish-American war, Cuba was under an American military government and it was natural for a US Army Commission on yellow fever to choose Havana for its work. The direction of their studies was largely determined by the earlier work of a Cuban physician, Dr Carlos Finlay y Barrés, who since 1881 had been convinced from his own clinical and epidemiological experience that yellow fever was a mosquito-borne disease. Finlay in fact provided the mosquito eggs from which the Commission's first breeding stocks of mosquitoes were reared.

Investigation of yellow fever in those days was a heroic business. No animal was known to be susceptible to the disease, and all experiments had to be carried out on human volunteers. They tried to work with milder strains of the disease, and most of the volunteers came through safely, but three paid the final penalty. The Commission proved three things conclusively: first, that yellow fever was carried by one particular type of mosquito (*Aedes aegypti*); secondly, that the patient's blood could infect the mosquito only during the first three or four days of illness; and thirdly, that the mosquito was not capable of infecting other human beings until ten or twelve days after its feeding on yellow fever blood. This knowledge was sufficient to show how to eliminate yellow fever from any city or town. The task was made simpler by the habits of the mosquito involved. It is a house-haunting mosquito that seldom flies away from any dwelling where it has once fed. By eliminating the breeding places of the mosquitoes, by shielding yellow fever patients from their bites, and by rigid measures to destroy at once all mosquitoes in any dwelling where a patient had become infected, yellow fever could be eliminated. In practice, yellow fever is the easiest of all diseases to deal with by public health measures, and its elimination from the West Indies and the more accessible regions of Central America was accomplished in an extraordinarily short space of time. All this was done without any knowledge as to the nature of the 'germ' responsible for the disease beyond some preliminary evidence that it was a 'filter-passing virus' and not a bacterium.

Definitive proof that a virus was involved had to wait until 1929. Yellow fever had shown increased activity on the west coast of Africa, and a group of American and English scientists, financed by the Rockefeller Foundation, undertook a study of the outbreak. Their first task was to seek an animal which was susceptible to the disease and for the first time the Indian rhesus monkey was tested. It proved highly susceptible, and with this discovery it became possible to use animals instead of human volunteers for yellow fever experiments. There was still danger enough in the work, however, and Stokes, the English member of the group, died of the disease just after their great discovery was made. It was soon established that the disease was due to a virus, not to a spirochaete or a bacterium, both of which types of microorganism had been incriminated by other investigators. Noguchi, a famous Japanese bacteriologist at the Rockefeller Institute, had some years previously announced that a spirochaete was the cause of yellow fever. The new work undermined his conclusions completely, and Noguchi also came to

West Africa to retest his theories. It is one of the great ironies of science that both Noguchi and his collaborator died of yellow fever soon after they had started their investigations in the Gold Coast.

The West African period, 1927–30, was the end of the heroic days of yellow fever investigation. Nowadays all workers with the virus have been artificially immunized, and are as resistant to the disease as a West African negro. Yellow fever research is now no more dangerous to those who undertake it than any other branch of scientific investigation. With the discovery that the disease was due to a virus capable of infecting monkeys, the Rockefeller Foundation initiated a large-scale international investigation of yellow fever. Soon after this was begun a method of producing the disease in mice was discovered, and with this cheap and convenient animal available, work on an even larger scale became possible. Since then all the major problems of yellow fever have been solved in principle, and great steps taken towards its control.

We may first briefly review the processes of infection, recovery and immunity in yellow fever. The virus is injected into the blood by the bite of an infected mosquito and sets up an infection primarily in the liver. The virus multiplies enormously, and at the height of the disease a single drop of blood contains many millions of virus particles. This flood of foreign material in the blood provides a very urgent stimulus to the antibody-producing mechanism, and if the patient is to recover, antibody appears within three or four days from the beginning of the fever. This is not enough at first to get rid of the virus from the blood, but it is sufficient to prevent infection of any yellow fever mosquito which may bite the patient. When recovery is complete, the blood contains large amounts of antibody, and this persists in only very slowly diminishing amount throughout the life of the person. Second attacks of yellow fever are practically unheard of.

If an individual has once been infected with the yellow fever virus his blood will always contain antibody. This fact, as with most other systemic viral infections, provides the epidemiological investigator with a simple means of screening populations for evidence of prior infection. By the 1930s virologists had devised a simple laboratory test requiring the inoculation of half a dozen mice. Large-scale surveys were undertaken over most of the world to determine where people were being infected with yellow fever. It was soon found that in addition to people who knew that they had had yellow fever a great many others possessed antibody in their blood who had certainly never had a recognizable attack of the disease. All the positive bloods, however, came from two well-defined regions: tropical Africa south of the Sahara; and Central America and the Amazon basin in South America. Unlike the other great plagues of the tropics yellow fever, as far as we can tell, has for some inexplicable reason never existed in Asia despite the presence there of the principal vector, *Aedes aegypti*. Largely as a result of these surveys there emerged an understanding of what, from our point of view, are two of the most interesting features of yellow fever, the frequent occurrence of infection without symptoms and the existence of jungle yellow fever.

Experience in Cuba, Panama and the coastal cities of Brazil had shown that it was easy enough to rid such areas of classical yellow fever by concentrating an attack on the *Aedes* mosquito. The growing hope that the disease could be finally eradicated from the world, however, received a check when in 1932 there occurred a disconcerting epidemic in a country district of Brazil where no *Aedes* mosquitoes were concerned. In the next year or two isolated cases or small epidemics of what came to be called 'jungle yellow fever' were detected in widely separated parts of tropical America. The same virus is involved but the epidemiology of jungle yellow fever is wholly different from that of the classical disease of tropical ports. The virus is maintained in a cycle involving jungle monkeys, marmosets and howler monkeys particularly, and mosquitoes that feed on the animals of the tree-tops. A very similar situation occurs in the rain-forests of central Africa. In both countries there is an inexhaustible reservoir of yellow fever unrelated to human habitations and not amenable to attempts at eradication. The jungle disease in South America most commonly affects timber cutters and others working in the forests, but if someone infected in the jungle falls ill in a town where there are *Aedes* mosquitoes an epidemic of 'urban yellow fever' may be initiated there.

In Africa the most puzzling feature of yellow fever is the vast extent of country in which antibodies indicating past infection are very common but the clinical disease extremely rare or unknown. It seems that the negro races have some innate resistance to the virus, but possibly more important is the relative insusceptibility of children. As in so many other diseases, first infection in childhood is usually a mild fever with no distinguishing features which lays down a lasting immunity. Through all its history yellow fever has spared those born in the regions of its greatest prevalence. It is significant that the only recent major epidemics in Africa, the 1943 epidemic in the Nubian mountains and the devastating epidemic in Ethiopia in 1961, occurred in regions at the edge of the area in which yellow fever is endemic. In the Ethiopian outbreak of what was essentially urban yellow fever a new vector *Aedes simpsoni* transmitted the virus from man to man.

The second phase of yellow fever control has been to devise means of protecting people from infection by jungle mosquitoes and of ensuring that no secondary epidemics spread by *Aedes* can develop. The first requirement is knowledge of the local situation. Are people being infected with the virus, are some dying of yellow fever without a definite diagnosis being made? Tests of blood, particularly of samples taken from children, will usually supply the needed information as to the prevalence of the virus. In South America very valuable information is obtained by the use of the viscerotome. This is a specially designed instrument by which an unskilled orderly can take a specimen of liver from the body of any person dying of a short-lasting illness. If death was from yellow fever, changes in the liver will be visible on appropriate examination under the microscope.

In any area where the disease is endemic the essential control measures are first to immunize the population and secondly to eradicate *Aedes* mosquitoes from the district by means fundamentally the same as are used in malaria eradication campaigns. In the early days, it was found adequate to reduce the number of mosquitoes to say 1 per cent of their normal abundance. That would soon stop any further cases of yellow fever and in the absence of any jungle reservoir of the virus it did not seem to matter much if the mosquitoes gradually regained their former numbers. With new methods of dealing with mosquito breeding, the present-day aim is nothing less than the elimination of *Aedes* from all urban areas in the two Americas. This has by no means been achieved but the extent of the reduction has been sufficient to prevent any cases of urban yellow fever since 1942. There is no practical approach to dealing with jungle yellow fever apart from the immunization of all who live or work in jungle areas. There was evidence in the last decade that the virus was moving slowly northward in Central America. As long as there are jungles with monkeys and mosquitoes in tropical America the control of *Aedes* and widespread human immunization will have to remain routine.

Immunization against yellow fever makes use of a strain of living virus called 17D which, on subcutaneous injection, gives a wholly symptomless immunizing infection. The principle is quite similar to Jenner's vaccination against smallpox; in fact, there is a very practical method used in West Africa of vaccination simultaneously against smallpox and yellow fever with a mixture of vaccinia and attenuated yellow fever virus. It was found early in the experimental study of yellow fever that virus grown in the mouse brain and passed repeatedly from

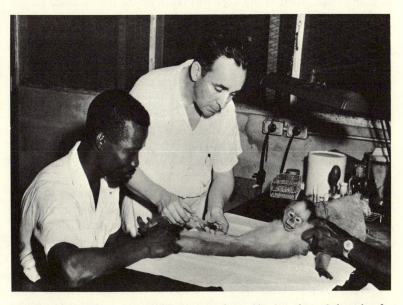

Fig. 30. (a) Yellow fever research in the Amazon forest: blood specimens being taken for antibody tests. (b) Yellow fever research in Trinidad: taking a blood sample from a *Cebus* monkey. (Both plates by courtesy of WHO.)

mouse to mouse gradually lost its power to produce the fatal liver disease in monkeys. By transfer through tissue cultures and then growth in chick embryos an even less virulent mutant was selected and in this way the immunizing strain 17D was produced. The vaccine was progressively improved, and its use has been gradually extended to larger and larger groups of people. The complication of 'serum jaundice' that marred its wide-scale use by the American Army in 1942 has already been mentioned. That difficulty has long since been eliminated, and the vaccine is used very extensively in infected areas. By international agreement yellow fever immunization is now compulsory for all persons travelling through the endemic regions of Africa and America.

It seems appropriate that the last of the historic plagues to be discussed should be the one which provides the outstanding example of the successful application of the ecological approach to infectious disease that has been the central theme running through this book. The story of yellow fever is an illuminating example of the effectiveness of biological research when it is supported by adequate finance, skilled direction and the goodwill of peoples and governments. By far the greatest share in the planning and support of the work was taken by the International Health Division of the Rockefeller Foundation. There were important incidental results from those years of work on all the medical, ecological and social problems that bore on the prevention of yellow fever. What was specially noteworthy was the way success in dealing with so important a disease brought the power of the ecological approach to the notice of all biological scientists. One might almost say that it was this success which made ecology respectable.

The work, too, opened a new world of the viruses carried by mosquitoes, particularly in the tropics. There are now more than 200 named varieties of these arboviruses, many of which are rather closely related to yellow fever virus. The first to be isolated was that subsequently called Murray Valley encephalitis virus, in Australia in 1918, but all the others came after the isolation of yellow fever virus in 1928. Most were actually obtained as by-products of the search for yellow fever virus in Africa and tropical America. Some are responsible for highly lethal haemorrhagic fevers or encephalitis, others for painful but relatively harmless arthritis, like dengue ('break-bone fever'), but most are still 'viruses in search of a disease'. Each has its own quite characteristic and usually quite fascinating story – its own special ecological niche in some remote jungle hideout. Almost invariably its natural hosts are creatures of the jungles or tropical swamps, monkeys, rodents or birds, often covering a number of different species and not infrequently being equally at home in birds and mammals. Its vector may be a single species of mosquito or a group of similar species, sometimes a tick and occasionally a sandfly or gnat. When man enters one of these ecosystems which have been established over millennia of evolution he does so at his own risk. Many of the tropical fevers that soldier, explorer or even the more intrepid tourist may contract are due to arboviruses, but they are quite irrelevant to the normal life history of the virus concerned.

20. Hepatitis, kuru and slow viruses

Many affections of the liver and bile passages can cause jaundice, but from the time of Hippocrates the occurrence of rather acute attacks of jaundice in young people has been well recognized. Jaundice has been particularly associated with military campaigns for many centuries, World War II and the Vietnam campaign being no exceptions. In the early years of this century it was labelled acute catarrhal jaundice and was not regarded as an infectious disease. Gradually the infectious nature of the condition became clear and the uncertainty about its incubation period and mode of spread was largely removed by the epidemiological studies of a country practitioner in Yorkshire, W. N. Pickles, who for the first time, in England at least, showed in 1928 that the incubation period was approximately one month and that transmission was by some form of person-to-person transfer.

Alone amongst major infectious diseases, catarrhal jaundice, now known as infectious hepatitis, has become progressively more frequent throughout the affluent countries of the world. It has never been conspicuous in populations of tropical countries but this does not mean that it is absent there. In fact young adults from Europe, America or Australia, working or soldiering for a year or so in tropical countries like New Guinea or Vietnam, have a very high probability of contracting hepatitis. The epidemiology of infectious hepatitis today shows an obvious resemblance to that of poliomyelitis before it was virtually eliminated by immunization.

In 1971 laboratories throughout the world are poised for the isolation of the virus of infectious hepatitis (IH), the last important agent of human infectious disease yet to be identified. Until now our knowledge of the disease has been built up largely at the clinical and epidemiological level. Limited studies on human volunteers during the 1940s played an important part, but there were too many deaths during the work to justify its continuance. Subject to more detailed confirmation when laboratory methods become available, probably within the next year or two, the following interpretation of the epidemiology seems to be well established. Basically IH is an intestinal infection like polio, the virus being taken in by the mouth and leaving the body with the faeces. In infants and very young children infection is almost symptomless. There is usually no jaundice and little or no fever but free excretion of the virus. There is a revealing story from a Pennsylvania orphanage in which there was a rapid turnover of teenage girls as nursing aides. The population of infants was also changing continuously as the children were taken for adoption or transferred to other institutions. There were the inevitable minor fevers and other infections among the children but jaundice was never seen. Yet the majority of the young nurses came down with typical infectious hepatitis within a month or two of commencing duty. The obvious, and we believe correct, deduction was that there was a persisting epidemic amongst the infants, kept going by the continual introduction of new susceptibles but showing no diagnostic signs. The nursing aides, most of them non-immune, were infected from the babies, but being older showed the classical symptoms.

In polio infections the great majority of children infected under the age of five show no signs of paralysis, while adolescents and young adults without immunity are more liable to severe paralysis. Much the same seems to hold in hepatitis, except that here the virus carried in the blood infects the liver cells instead of the cells of the spinal cord. This at once clarifies the epidemiology. In crowded communities with poor hygiene oral infection occurs very early in life without symptoms and with subsequent immunity. In the Yorkshire countryside there is no endemic infection but occasional epidemics of hepatitis move from family to family by personal contact, almost like measles except for an incubation period more than twice as long. In the urban communities of the West each improvement in general sanitation will raise the average age of first infection and, as was the case with polio, an invisible endemic affection of infants and toddlers is steadily changing its character to become a clinically evident epidemic disease mainly involving school-age children.

Everything suggests that considerable amounts of virus are present in faeces and that only a small number of viable particles are needed for the infection of a susceptible child. It seems likely that the commonest type of person-to-person transfer is by minor faecal contamination of the hands. This naturally does not preclude infection by grosser types of pollution by sewage. The biggest epidemic on record took place in New Delhi in 1955 when water supply channels were contaminated from sewage pipes broken during a flood. The danger was recognized immediately and heavy additional chlorination applied to all suspect water. The action was apparently effective against typhoid and dysentery bacilli but not against the hepatitis virus, for some 1100 cases of the clinical disease were reported over the relevant period. Oysters grown in sewage-polluted estuaries may also convey the disease. A small epidemic from this source in Stockholm differed from almost every other epidemic on record in involving only wealthy men – oysters being a luxury in Sweden.

Search for means by which the virus could be recognized or cultured has been prolonged and intensive. There have been many candidates but only in the last five years has a real lead emerged. But it came from the study of a different disease, 'serum hepatitis', and is best discussed in relation to that condition.

*

Anyone concerned with the treatment of syphilis in the 1920s knew of the condition known as salvarsan jaundice which appeared rather irregularly amongst patients on extended courses of intravenous injections of one or other of the organic arsenicals then being used for treatment. It was regarded as resulting from a toxic effect of arsenic on the liver and it was many years before the real cause was recognized. In fact 'salvarsan jaundice' was serum hepatitis, resulting from the accidental transfer of virus from one patient to the next individual injected with the same syringe. Similar things tended to happen wherever large numbers of people were being injected or subjected to other procedures involving taking or administration of blood. The fact that serum from apparently normal people could produce the disease was shown in 1938 by the occurrence of hepatitis in children given convalescent serum as a protection against measles. Universal interest in and recognition of the nature of serum hepatitis dates from the American Army's experience at the outbreak of the Pacific War following Pearl Harbor. As outlined earlier, there were 28,000 cases of hepatitis with some sixty-two deaths in American servicemen who had been immunized with certain batches of yellow fever vaccine. It was a live virus vaccine which had been found to retain its effectiveness well if made up with normal human serum. The human serum was used instead of some animal serum to avoid any possibility of anaphylactic sensitization against a foreign protein, but it was soon incriminated as the cause of the hepatitis.

Once it became clear that a proportion of apparently normal people – and by no means an insignificant proportion in the United States – had something in their blood which produced hepatitis when injected into others, many things previously unexplained became clear. One of them was the high proportion of patients who two to three months after a blood transfusion had an attack of jaundice. People began to speak of icterogenic (jaundice-producing) blood or plasma and an occasional blood donor was identified as a menace in this regard. In the later stages of the war in Europe hepatitis was rife amongst American troops. Some was clearly infectious hepatitis, some almost certainly serum hepatitis; most of the cases could have been either. Clinically the two conditions were indistinguishable and the only way of distinguishing one from the other in the absence of laboratory tests was according to the length of the incubation period. Serum hepatitis (SH) usually came on two to four months after transmission by some form of injection, whereas the incubation period of the epidemic form (IH) was about one month. Under conditions of active service, however, details of the time and mechanism of infection were virtually never available.

Over the years since the war there has been active research on all aspects of the two forms of hepatitis with the answer always seeming to be just around the corner but never quite materializing. Epidemiological studies and limited use of human volunteers led to certain substantial conclusions. First, that most, perhaps all, of these blood donors whose blood produced SH were chronic carriers of the virus, showing no symptoms. Second, that the prevalence of carriers in the community was directly related to the availability of medical care, presumably reflecting the frequency of use of injections under the skin or into veins and the removal of samples of blood from an arm vein or by finger prick. Third, that hepatitis was very frequently seen in heroin addicts and would presumably also occur in any other type of drug addiction involving injection of the dose. Whether for these reasons or not, there seems to be no doubt that there is a much higher likelihood of being infected with SH by a blood transfusion in the United States than in any other country.

Quite obviously there has been for years outstanding need for two developments, (1) a means of knowing whether a given donor's blood contains SH virus before it is used for transfusion, and (2) a means of immunization against IH.

As has rather often been the case, the discovery that has led to what seems likely within a few years to fulfil these needs came from a chance finding in another field. In a programme of study on protein antigens present in serum of different human races, Blumberg tested an Australian aborigine's serum against a reagent commonly used in such work – the serum of a patient who because of some form of chronic anaemia had been given many blood transfusions. Such serums are liable to contain multiple antibodies against any proteins different from their own which happened to be present in the blood they had been given. When the Australian aborigine serum was tested, it gave a reaction of mutual precipitation with this particular antiserum. Since the reaction was not seen with any of the other serums that were available to Blumberg he assumed that a new substance which might be characteristic of Australian aborigines was present in that serum, so he called it Australia antigen or Au. This was in 1967. A subsequent search for other sources of Au using the same test antiserum gave results which made it clear that Au antigen was not a racial characteristic. It occurred in a number of American hospital patients and soon it emerged that it was in some way associated with serum hepatitis. Further, when the Au antigen was isolated from the serum, it was found not to be a soluble protein but to be made up of tiny units which by electron microscopy had all the appearance of virus particles. There is now general agreement that the Au particles are directly related to the SH virus but no one has yet shown that the virus can be grown in cell culture or in any laboratory animal.

One or two new ways of detecting Au antigen in the laboratory have already been developed and there is no doubt in anyone's mind that by the time this book is published a routine test of blood donors for Australia antigen will be standard practice in blood transfusion services. It promises to be the first important new life-saving procedure of the 1970s.

There are now strong indications that IH virus is similar to that of SH in a number of ways. Faeces from cases of IH contain particles which resemble Au antigen in the electron microscope but show different immunological qualities. Not much has yet been published in this field but the story rings true. There is also recent laboratory gossip about promising efforts to grow IH virus in cell culture. Repeated false alarms over many years have made virologists sceptical about 'hepatitis breakthroughs' but with two approaches to provide mutual help the prospect this time is better than ever before.

There is no doubt about the desirability of an effective vaccine against hepatitis, but opinions differ as to its potentialities and it is probably best to underline the fact that no predictions are really possible until the IH virus can be grown regularly in cell culture and its properties fully investigated. If like most virologists we envisage a live virus vaccine we can hope that it will produce almost as substantial an immunity as follows natural infection. On the other hand it may turn out that any degree of attenuation will diminish the antigenicity of the vaccine below the effective level.

Serum hepatitis has its special character because the virus is a 'poor antigen' in the sense that it is much more likely to provoke tolerance that allows it to persist in the serum for years or perhaps indefinitely, rather than a brisk immune response that will eliminate it and leave the individual immune. Neither is IH a really good antigen, though much better than SH. In a typical case the incubation period is about one month and the virus may remain in the blood for up to another month. It would only need a relatively minor mutation of the virus to turn IH into something with the same sort of behaviour as SH.

This leads us to a final problem which we have not so far mentioned. Where did serum hepatitis come from? It is today a creature of the hypodermic syringe and the intravenous injection, but until the turn of the century injections were rare and the 'ancestor' of serum hepatitis must have had other ways of surviving. We shall probably be in a better position to speculate in a year or two, but in 1971 the most likely answer is that SH is a derivative of one of the IH viruses which have been moving around human communities since urban aggregations first appeared. Indeed, the present SH virus can be transferred on occasion by the faecal–oral route, and the IH virus by serum. An IH virus has only to do what other viruses readily do in the laboratory and it is virtually an SH virus. It must merely undergo a minor antigenic change and lose some of its infectiousness, in this case its capacity to pass readily from one host to the next by the faecal–oral route.

There is a real possibility of the existence of a range of related forms of these hepatitis viruses with varying potentialities of producing disease. In 1947 there was an almost epidemic prevalence in Sweden of fatal hepatitis involving women in their forties and fifties. In the Solomon Islands there is an exceptionally high incidence of people with Au antigen – 30 per cent of nearly 1800 individuals tested. Hardly any of these people give any history of being jaundiced and few would ever have had injections of any sort. The main clue to its origin came from an elaborate analysis of family relationships of those with and without the antigen, which pointed to a genetic factor being concerned. According to this interpretation, the presence of an autosomal recessive gene was needed for an infection by some natural route to take on the quality of persisting tolerated infection which alone can give a positive Au test. In other words a hepatitis virus originally spread by the faecal–oral route finds it very easy, in an individual of the right genetic group, to persist in the serum for life.

Findings of this sort are raising expectations that a whole new series of 'slow viruses' may soon be brought to light. The only example known

255

to human medicine so far is the virus isolated from another Melanesian condition, 'kuru'.

*

Of all the infectious diseases of man kuru is perhaps the most extraordinary. It is confined to a few hundred square miles in the Eastern Highlands of New Guinea. It seems to have been unknown before about 1910, then gradually increased in incidence until in the early 1950s it was by far the most frequent cause of death in adult women. It was almost certainly spread by ritual cannibalism and its incubation period is up to twenty years. With the disappearance of cannibalism around 1954 the disease is now dwindling fast. The partial elucidation of the origin of kuru by Gajdusek and various collaborators over the years since 1957 represents such an outstanding contribution to the understanding of infectious disease that it justifies our telling the story at some length.

Kuru had been mentioned by patrol officers and anthropologists previously but the first medical observation of the disease was in 1956 when a medical officer visited the Okapa area about fifty miles east of Goroka in the New Guinea Highlands. He noted the general character of the disease and its high incidence in the area. During the following year Gajdusek began detailed studies of kuru and the region has been under medical surveillance ever since.

The situation has changed over the years and it is convenient to describe the classic picture of kuru as it appeared in the 1957–60 period. Kuru then was widespread amongst women and children of a single linguistic group, the South Fore, indigenous to a well defined area of the Eastern Highlands and numbering about 15,000. The disease was invariably fatal with a duration from first symptom to death of six to twelve months. The typical sequence in a young woman would begin with her noticing that she was clumsy with her fingers in such tasks as rolling hair into string. At this early stage the only clinical sign was some tremor of the hands. The incoordination of movements grew slowly worse and took on the form characteristic of a progressive cerebellar degeneration. Eventually it became impossible for her to walk, owing to complete incoordination, although some muscular power remained to the end. The last stage was one of complete incapacitation often with inability to swallow. The patient was cared for by her female relatives inside the hut and was quite unable to move herself. Death resulted from the inevitable complications, starvation, inhalation pneumonia, infected burns and bed sores. Post-mortem the only significant findings were degeneration and cellular proliferation in the cerebellum and some other regions of the brain.

Apart from an occasional instance of hysterical mimicry of kuru symptoms the condition was unmistakable and invariably ended in death. In the most heavily involved villages nearly half the deaths of adult females and most of the deaths of children of both sexes between the ages of five and sixteen were due to kuru. Kuru in fact completely dominated the social life of the people, causing a serious shortage of women. At a village one of us visited in 1962 the head man, the luluai, had lost four wives with kuru; lesser men had to do without.

The area came under Australian control in 1954 and one of the main difficulties of the early patrol officers was to stamp out the custom which required the ritual murder of any man (usually from another village) who had been recognized by appropriate magic as the sorcerer who was maliciously harming a man by bewitching his wife with kuru. This 'pay-back' murder had some particularly gruesome features and had become almost as dominating a cause of death amongst adult males as kuru was amongst the females. Apart from an occasional case in a young man, kuru was never seen in adult males but it was approximately equally common in male and female children. The symptoms in children were essentially similar to those seen in women and death was equally regular.

The disease was confined to the South Fore people and to those adjacent linguistic groups with which there was occasional inter-marriage. No outsider, European or Papuan, coming to the region contracted kuru and the only cases observed outside the area were in two young men from the Fore area who had come down to a coastal area as labourers.

The most likely interpretation seemed to be that this was a genetic disease and it was suggested that the situation might be explained if a mutant gene K was dominant in females but recessive in males. On this view homozygotes (KK) of either sex developed kuru in childhood while the heterozygotes (Kk) showed late onset of the disease in women but no symptoms in males. In view of the possibility that perhaps 40 per cent of healthy males were, on this view, heterozygotes who could theoretically introduce the gene into previously uninvolved areas there was a move to prohibit recruiting of labour from the area. For a few months there was in fact what we believe was the only attempt ever made to 'quarantine' a dangerous human gene. It proved quite impractical to police the embargo and the scheme was quietly dropped.

Everyone was aware that the genetic hypothesis had one fundamental weakness. It demanded that a dominant or partly dominant mutation which must have arisen in a single individual perhaps a thousand years previously had had such a striking survival advantage that by 1950 it was present in several thousand individuals. Yet the gene was known to be highly lethal.

A possible way around the difficulty was found when close questioning of old men in the course of establishing pedigrees and the like, made it clear that kuru had come to the area about 1910 and had moved slowly around the Fore territory, only reaching the currently most active area around 1940. This meant that the genetic theory might still be correct if the kuru gene had no harmful (and perhaps a positively beneficial) effect under normal conditions, but that in the presence of some environmental factor it became lethal. There was the analogy of glucose-6-phosphate-dehydrogenase (G6PD) deficiency which predisposed many negro servicemen to suffer from haemolytic anaemia when they took antimalarials.

At the time the favourite suggestions for the environmental agent entering the Fore region about 1910 were a new food plant, with maize as a possibility, or a virus disease, perhaps of European origin. Heavy metal poisoning was also considered. On the whole the most likely hypothesis seemed to be that a virus acting against an unusual genetic background was responsible.

Then rather precipitately the whole genetic hypothesis was abandoned when three important new facts were recognized. (1) Children were no longer going down with kuru. (2) Anthropologists found that cannibalism was first introduced to the South Fore people around the critical time of 1910. (3) A disease with the characteristics of kuru was produced in chimpanzees inoculated with brain material obtained from victims of kuru.

Each of these points justifies expansion. (1) From about 1954 regular patrols throughout the area ensured that the 'village book' should be kept up to date with entries of all births and deaths. In all medical surveys, the age of anyone born before 1954 had to be guessed but that of children born since could be verified. In 1957–9 there were many children with kuru, and 117 deaths of children estimated to be between five and ten years of age were recorded. The number has fallen progressively, only twenty-three such children dying in 1961–3, and in 1970 no new cases in children were seen. Adult cases still occur but the number is decreasing and it is noteworthy that their average age at onset is rising.

(2) According to Robert and Shirley Glasse cannibalism was not a traditional custom of the South Fore people but was introduced in the

period 1910–20 by women who had learnt the habit from women of neighbouring tribes. The Fore men were not interested. The senior author has a vivid memory of an afternoon in the Glasse's native-built house in Wanitabi in the very heart of the kuru country when Bob Glasse was telling a group of medical visitors his conclusions about cannibalism. In his view it came in as a vicious practice amongst the women and was soon ritualized as a social necessity to eat some of one's dead kinswomen's brains. The men however were conservative. Cannibalism had never been a habit with their people and they resisted the idea – in particular they were convinced that 'eating women was wrong' and possibly dangerous. Glasse felt then that the simplest explanation of kuru incidence was that a virus was taken in with semi-cooked brain from a deceased victim. The women consumed most of the material but titbits went to the children including pre-pubertal boys. The men stayed aloof and avoided kuru.

As soon as the Australian administration took over the area in the period 1954–6 cannibalism was severely frowned on and rapidly disappeared. Soon it became more usual for a brain to be packed in dry ice for transport to Bethesda than to be eaten by sorrowing female relations!

(3) There is nothing equivocal about the results with chimpanzees inoculated into the brain with kuru brain substance. Eight different human brains have each produced a kuru-like disease in chimpanzees with incubation periods of eighteen to twenty-four months. Passage has been successful to other chimpanzees and to spider monkeys. Work as yet unpublished indicates that the agent responsible is filterable and presumably a virus.

So in 1971 virtually all those interested in kuru accept the view that a necessary part of the process by which kuru develops in a Fore woman is that she should be infected with the kuru virus by handling or eating the brain of a previous victim. The most convincing evidence is the recent progressive disappearance of kuru in children which fits perfectly with the timing of the elimination of cannibalism as a ritual. Certainly no straightforward alternative explanation has been suggested but even if we accept cannibalism as an essential part of the story, there are still many aspects that require elucidation. Women and young men who have certainly not indulged in ritual cannibalism for more than twelve to fifteen years are still in 1971 coming down with kuru. Irrespective of their age when the first symptoms appear, death results within the year. It seems incontrovertible that the virus must have set up a chronic persisting infection which is symptomless for a long and variable period of years. Then at some apparently arbitrary moment a new factor

259

initiates the process which ends with lethal infection of the brain. The incrimination of a virus provides no real answer to the question of what trigger sets the kuru process in action and what is the pathology of that process which leads so regularly to death in about nine months. There are other questions to be answered. Where did the kuru virus come from and how did it survive as a virus before 1910? Has the possibility that there is a genetic susceptibility of women in this general area of the Highlands been eliminated? What is the significance of the fact that brains from patients in England and America with the rare Creutz-feldt–Jacob disease, which has some clinical resemblance to kuru, have produced analogous disease on inoculation into chimpanzees? They certainly were not cannibals. Finally what is the relation if any of kuru to scrapie in sheep? This last question merits brief discussion.

Scrapie is the Scottish name for a disease of sheep that has been known in Europe for at least a century. The name is derived from the habit of affected sheep rubbing themselves against a fence post or anything else firm enough, until much of their wool is scraped off. It is a disease limited under natural conditions to certain well defined genetic strains of sheep and according to field investigators it 'spreads' as a typical recessive character without ever showing evidence of sheep-to-sheep infection. Symptoms of scrapie appear at any time from two to five years of age and in addition to the itch, which presumably represents damage to the sensory system, there is a progressive cerebellar degeneration giving symptoms broadly equivalent to those of kuru. The pathological lesions in the brain are closely similar and, as in kuru, a virus, or at least a transmissible agent, can be isolated from the brain of an affected sheep. The experimental disease produced by intracerebral injection of brain material in sheep, goats and mice has an incubation period of several months.

Scrapie virus has not yet been clearly visualized with the electron microscope and it is quite extraordinarily resistant to damage by heat or formalin. No antibodies have been detected in affected animals. Strictly speaking, therefore, the infectious agent has not yet been proved to be a virus, nor for that matter has the kuru agent.

The evidence that naturally contracted scrapie is limited to sheep of a particular genetic constitution makes us reluctant to discard the possibility that genetic factors may be concerned in susceptibility to kuru. Serum hepatitis is one human disease with analogies to the type of persisting infection which must be postulated for kuru. In the earlier part of this chapter we described how in another region of Melanesia not far from New Guinea. Au antigen indicative of persisting serum hepatitis is found only in persons of an appropriate genetic constitution. There, it seems that the capacity to become a carrier following infection is transmitted as an autosomal recessive. Something of the same sort may yet be found to play a part in the epidemiology of kuru. So far there are only speculative answers to any of these questions. Clearly there is plenty of scope for fruitful investigation of both serum hepatitis and kuru in the future.

We have written about these two diseases in a single chapter because of their central common quality of a long period of persisting symptomless infection. The connection may turn out to be incomplete if future work demonstrates a clear distinction between 'chronic' infections (like serum hepatitis) in which the virus persists for years following an acute disease, and 'slow' infections (like kuru) in which the virus is present for years before the disease becomes manifest. One cannot however escape the opinion moving around the laboratories that kuru and serum hepatitis have real analogies, and may become prototypes for a whole range of human illnesses whose cause is at present unknown. We sense the beginnings of a new medical science with at present only an informal name – 'slow virology'. None of these diseases is going to be easy to investigate or understand. At least three major disciplines, virology, genetics and immunology, will need to cooperate for their elucidation and we would not dare to predict that understanding will necessarily bring ways of practical control. Whatever the outcome, here is the major intellectual challenge of infectious disease for the final decades of this century.

21. Perils and possibilities: an epilogue

As long ago as 1937 when the first edition of this book was being written microbiologists were becoming worried at some of the social implications of their work. Some were already aware that each step in the prevention of the lethal diseases of the tropics would be equally a step toward a population crisis. Others were concerned about the potentialities for the use of microorganisms of disease as weapons for mass destruction in war. Both topics have been discussed along with a variety of speculations on the future of infectious disease in the final chapter of each edition. This has made it an interesting exercise to revise the book three times, approximately in 1950, 1960 and 1970, and to assess the changes in each decade in the understanding and control of infectious disease and also of course to reconsider what happened in regard to population growth and biological warfare in each interval.

Between 1940 and 1950 there was a wealth of significant change. It was a decade that saw the development and use of the antibiotics and the synthetic insecticides. The Second World War was less influenced by infectious disease than any war in history. Serum hepatitis appeared on a relatively large scale as an iatrogenic disease, and armies intruding into alien ecosystems suffered from malaria, scrub typhus, and other infectious diseases, but far less than in similar campaigns of the past. Between 1950 and 1960 a new pandemic of influenza had curiously little impact compared to that of 1918–19. The Salk and Sabin polio vaccines were developed and used with resounding success. Cell culture methods brought dozens of more or less trivial viral infections to light and shortly before 1960 the first of the myriad viruses of the common cold had been cultured. The technical achievements of the 1940–50 period were refined and more effectively applied and the process of reducing the mortality from childhood infections continued in gratifying fashion. The results of chemotherapy of tuberculosis became evident in most countries just after 1950 and probably represent the highpoint of achievement in that decade. Between 1960 and 1970 the results went on improving and in the research laboratories, greatly improved facilities, a better educated and constantly growing work force, and the new concepts of molecular biology, produced a flood of scholarly work. Yet in regard to the natural history of infectious disease in human populations nothing has greatly changed. Some important changes for the worse are not Nature's doing. Venereal disease, which seemed to be vanishing in the 1950s, is now on the increase amongst young people in all Western countries. As noted earlier, this may well be one of the less desirable results of the availability of contraceptive pills. Fortunately there is no evidence that penicillin-resistant strains of the spirochaete of syphilis have yet emerged and although the gonococci became resistant to sulphonamides in the 1940s antibiotics are still effective. In other types of infection drug-resistance continued to be troublesome but the introduction and use of the semisynthetic penicillins was an important counter measure. Another important change for the worse was the emergence of malarial parasites resistant to most of the synthetic drugs. This has for obvious reasons been most evident in Vietnam. On the positive side the development of methods for culturing and vaccinating against rubella virus and the recognition of the Au antigen in serum hepatitis are the main landmarks.

On the basis of what has happened in the last thirty years, can we forecast any likely developments for the '70s? If for the present we retain a basic optimism and assume no major catastrophes occur and that any wars are kept at the 'brush fire' level, the most likely forecast about the future of infectious disease is that it will be very dull. There may be some wholly unexpected emergence of a new and dangerous infectious disease, but nothing of the sort has marked the last fifty years. There have been isolated outbreaks of fatal infections derived from exotic animals as in the instance of the laboratory workers struck down with the Marburg virus from African monkeys and the cases of severe haemorrhagic fever due to Lassa virus infection in Nigeria. Similar episodes will doubtless occur in the future but they will presumably be safely contained.

Preventive medicine is already turning its attention to the conventionally less serious childhood diseases, with vaccines against measles, rubella and mumps available and being used to immunize progressively larger numbers of children each year. Doubtless there will be pressure to extend the range of relatively minor infections against which some form of immunization can be offered. Opinions differ as to the likelihood of success but most public health workers would probably agree that vaccines against trachoma and against infectious hepatitis are outstanding needs. Some have suggested that immunization against syphilis and gonorrhoea is possible in principle and a vaccine against respiratory syncytial virus, the cause of the most serious respiratory infection of babies, would also fill a need. We can be confident that research will

continue and that vaccines which prove effective and have no un-
desirable complications will come into use.

There are two areas of current research which have possible significance for future practical use. These are both concerned with virus
disease – the use of specific chemotherapy for treatment, and non-
specific prevention by the induction of interferon. In general, older
workers find good reasons for being highly sceptical about the potentialities while younger ones feel that the possibilities are almost unlimited.
Events should show which are right within the next ten years.

*

This forecast assumes that in general things will continue as they are,
that there will be no major war and that the population explosion will be
contained. If things turn out differently the picture could change
disastrously. The control of infectious disease in heavily populated areas
depends on a functioning technological civilization. Any serious breakdown of that organization will bring famine and epidemics in a time
inversely proportional to the density of the population. We have neither
of us sufficient sense of history or political sophistication to do more than
reflect majority opinion that the overriding danger is large-scale nuclear
war, but as microbiologists we must be deeply concerned about the
immediate sequelae to a major nuclear conflict. To use nuclear weapons
for war is biologically absurd and socially insane. Nevertheless the
bombs exist and we must presume that some day they will be used.
Maybe the 'six hour war' will destroy most of the major cities of the
technologically developed world and kill perhaps a hundred million
people. For the other millions, lethally irradiated but taking weeks to die,
a large proportion will die of infection by otherwise harmless microorganisms as a result of partial or complete paralysis of the immune
response. What would develop subsequently is completely unpredictable.
It will obviously depend almost wholly on the degree to which public
order can be restored. If there are adequate undamaged stockpiles of
antibiotics, vaccines and other medical supplies, and if health and
hospital services can function more or less effectively, epidemics of
bacterial pneumonia or other lethal infections will probably be pre-
ventible. We might guess that if medical help is wholly inadequate
epidemics of the same type as caused most of the mortality in the
1918–19 influenza pandemic would develop rapidly. In a sense the
damaged immune system in irradiated people could be equated with
superficial damage to the lungs by virus infection. Either could allow
pneumococci, streptococci and staphylococci to take on epidemic

virulence which they could never exert in normal people. Even so, complete social disintegration among the survivors of a densely populated modern state would let loose so many other modes of untimely death that infectious disease might be rampant yet pass almost unnoticed.

Fear of bacteriological warfare on a major scale seems to have lessened in recent years but everything suggests that there is no technical impediment to its use as a strategic weapon. The barrier is primarily psychological. It is probably significant that all current discussions of military power are concerned with atomic weapons and the means of delivering them. Bacteriological warfare has been little discussed, public reaction against it is unanimous and both the USA and the USSR have voluntarily reduced their capacity in that field in a fashion that is as yet unthinkable in regard to nuclear weapons. It may be that the curious sensitivity of political and military leaders in this respect offers real hope that BW methods will never be used. The claim by North Koreans that biological warfare was being used against them in 1951–2 was probably baseless, but it had a very significant effect on public opinion in Asia and perhaps had an unconscious influence on Western thought. War is a traditional method of human behaviour but the use of poisons and microbes is not part of the tradition. There has always been war and wars have always up to the present been fought according to a traditional pattern constantly evolving but never showing any sharp discontinuity. There has always been a hierarchy of leadership and the inculcation of loyalty and discipline. Weapons have evolved continuously from axe and spear to sword and arrow, crossbow and catapult gave way to firearms, the horse and the war-canoe to battleship, tank and bombing plane. Ball and shrapnel were replaced by high explosive and the proximity fuse, and now the atomic bomb and the intercontinental ballistic missile bring the tradition up to date.

But the use of biological warfare is an absolute novelty, a complete break in the tradition of war as an extension of personal combat. As Vannevar Bush has pointed out, there is bound to be an intense resistance on the part of the statesmen and generals to the use of such a method, not necessarily from any moral feelings but simply because of the intrinsic human resistance to any violent break with an existing social pattern.

This may be over-optimistic. Power politics can attract psychopathic personalities not sensitive to the normal inhibitions of the great majority of people. It is public knowledge that until very recently active research on the military use of disease-producing bacteria, rickettsiae and viruses was underway in American and British establishments. Much of the work has provided results of great interest to bacteriologists generally and has been openly published in the technical journals. Other countries have not admitted the existence of bacteriological warfare research within their boundaries but obviously every military group in the world must be deeply concerned with the topic in one way or another. The published papers indicate that the anthrax bacillus and botulinus toxin were being closely studied in 1945–6, while some attention was also being paid to the organisms of tularaemia and psittacosis. Any experienced bacteriologist will know that there would be great technical difficulties in producing and using dangerous bacteria in war, but he would also realize that most of the difficulties are only technical ones. There are of course no open accounts of what forms of weapon have been developed. We can imagine a variety of possibilities but the essence of the matter is that it is physically possible to produce in a room a thin mist of bacteria so that any animal that takes a few breaths in that room will die unless it is treated subsequently with an appropriate drug. To produce similar conditions over the large volume of air within and around an enemy city is physically possible and in all probability the technical methods of achieving this have already been perfected.

A ton of anthrax spores would contain about 10^{18} individual spores. If these could be uniformly distributed in a volume of air six or seven miles across and extending 300 feet upwards from the ground each litre of air (about one deep breath) would contain about a hundred thousand spores. There are a number of microorganisms which induce fatal infection when less than ten individual organisms are introduced into a susceptible experimental animal. Those calculations in one sense are misleading. Nothing approaching such uniformity of distribution could be produced under the most ideal conditions. But they do suggest that even the very inefficient distribution over a city of the number of bacteria that could be carried in a single plane or a single cluster of bombs might produce a startlingly large toll of illness and death. The men with the computers who work out the arithmetic of mutual extermination by nuclear bombardment are no doubt well aware of the improved kill that could be obtained by the delivery of properly chosen microbiological weapons a day or two after a nuclear attack.

There is unease everywhere amongst research microbiologists with a social conscience. Sometimes we wonder what is the justification for pushing ever more deeply into the nature of viruses and other microorganisms of disease, when all the public health problems of infectious disease have been solved in principle. Will an understanding of the genetic control of virulence in poliovirus help us to prevent polio? The straight answer is an unequivocal No, yet it is an equally acceptable answer to say that accurate scholarly knowledge about any aspect of the universe is good in itself. The whole philosophy of our technological civilization is that any scientific discovery is potentially capable of being applied to human benefit. Such an application must be sought and usually some justification for its development will be found. It is in line with much current thought – by Lewis Mumford for instance – to believe that 'human benefit' in its current usage means little more than profit or power for someone. We are worried about the possibilities of misuse of much of current work on microbial genetics. Without doubt the published results have illuminated the whole of biology and such knowledge would be wholly desirable and harmless in a world where war and assassination were unthinkable. But as long as the tradition of the irresponsible use of power persists we must continue to fear that the steady advance of experimental microbiology will progressively increase the likelihood that effective biological weapons of mass destruction, sabotage and assassination can be produced.

This is a situation that must be faced. While war is possible, the development of microbiological weapons will go on and if they are perfected they will be used as seems expedient when war occurs. Apart from the fact that they represent a break in the traditional techniques of warfare there is no reason to consider them more inhuman than any other method of killing. Every legitimate argument against their use in war is equally an argument against the whole institution of war.

It cannot be too urgently stated that the advance of science is allowing the development of wholly unnatural modes of human domination and mass destruction – unnatural in the sense that they represent conditions never encountered in the course of human evolution. In the physical field no mammal prior to this century was ever exposed to penetrating radiation. Nor, except in small experiments in recent years, had any mammalian species to face the impact of a pathogenic microorganism that had not evolved and been able to survive by natural processes. Perhaps in another field something equally unnatural is emerging in the new techniques of forced public confession and 'brain washing'. Somehow the use for human domination of nuclear reactions, artificially virulent and artificially disseminated microorganisms and control of thought and behaviour by unnatural methods, whether pharmacological or psychological, must be prevented. Unless this can be done the whole biological and social background to the human species is in danger of complete disorganization of a type so different from any previous experience that nothing less than a new process of evolution will be needed before a healthy civilization again becomes possible.

*

In a brief final section we must turn again to the implications of infectious disease and its prevention on human populations. An implicit theme running through many of the preceding chapters has been that the prevention of death from infections in infancy and childhood has been the direct cause of the change from a virtually stable human population to the present explosive increase at the rate of something like 2 per cent per annum overall and up to 3·2 per cent in large areas of the tropics, notably Mexico and the Philippines. In the thirty years since this book was first published world population has increased from 2300 to 3700 million and the current rate of increase is faster than it has ever been. In Chapter 18 we cited the demographic experience of Mauritius as a model of what happened when endemic malaria was suddenly removed. Figures for Ceylon given by Cockburn show almost as sharp a change between 1945–6 and 1947–8 as occurred in Mauritius. The birth-rate rose (37·4 to 39·9), death-rate fell (21·0 to 13·7) and the rate of population increase per annum changed from 1·81 per cent to 2·62 per cent. Much of the accelerating population increase in tropical countries can probably be ascribed equally directly to the progressive reduction of malaria since the end of World War II. At the same time however sulphonamides and penicillin became freely available with a concomitant reduction in all those deaths for which bacterial infection was primarily or secondarily responsible.

Although changes since 1946 can be ascribed unequivocally to the progressive elimination of death from infection by the use of antibiotics and antimalarial measures, all the statistics show quite clearly that the disequilibration of birth and death levels began for Europe at least two centuries ago and the population increase was strikingly evident long before deliberate scientifically based action against infectious disease could have been of any significance. The general improvement of standard of living over the last two hundred years was the important factor in the early stages. Each improvement in nutrition and housing, comfort and cleanliness helped lower the mortality of childhood. The two factors have now joined forces. By the combination of a rising standard of living and deliberate action to prevent and cure infection we are moving rapidly to the stage when the expectation of life at birth is everywhere around seventy years. To achieve a stabilized population under those conditions will require birth- and death-rates around 14–15 per 1000 and about 18 per cent of the population under fifteen. Yet in the developing countries of the world during the 1960s birth-rates ranged from thirty-eight to forty-five and 43–46 per cent of the population was under fifteen.

To engineer the change from the present catastrophic imbalance to a stabilized global ecosystem will require a colossal human effort. Even the most optimistic demographers ask for more time than circumstances seem likely to allow us. There is probably no one who can even hazard a guess when or whether the only sane ecological climax for the human species with a world population between 1 and 2 billion will ever be attained. As things stand there seems no acceptable and technically feasible approach by which world population can be stabilized below 20 billion people, the number which will be reached in fifty to a hundred years' time.

Our specific interest in overpopulation is whether under modern circumstances the associated overcrowding will in itself cause a serious deterioration in the control of infectious disease. Again the answer seems to be that as long as social order can be maintained and public health given reasonably high priority in government spending there is no reason why any breakdown should occur. Epidemics of new type or exceptional intensity will probably come only as a result of serious disintegration of the social structure. Civil war and rising incidences of crime, drug addiction, psycho-somatic and mental disease, are likely to be much more important results of overpopulation than any increase in infectious disease.

No biologist can avoid the conviction that a potentially disastrous crisis of human population is in the making and that the most dangerous aspects of overpopulation will come from human conflict in various forms rather than from famine or infectious disease. It is an evolutionary necessity that a species must find an approximately stable population level in any ecosystem of which it is a part. In the long run this must hold for man. In one way or another, means must be found by which the population of a given region does not increase beyond the level that the region in question can support from its continuing resources. An equally important corollary is that the irreplaceable resources of the world must be conserved until effective substitutes are in sight that do not deplete capital resources. Conservation is sound ecology.

Finally, a little may be said about war and human conflict from the biological viewpoint. Most of the ostensible and some of the real causes of war lie in this field of interaction between population levels and natural resources, whether for food production or for the provision of raw material for industry. A broad ecological approach is the only answer.

The other side to human conflict is, as we all know, 'human nature', selfishness, aggression and desire for power on the one hand, loyalty to individuals, groups or ideas on the other. These things may seem to lie right outside of the province of science, but one of the most interesting developments of modern biology, the comparative study of animal behaviour, is making it clear that 'human nature' is by no means confined to human beings. It is not too much to say that all the basic elements of human conflict are visible in animal behaviour. In the last five years there has been a flood of popular and semipopular writing on the findings of the ethologists who have studied animal behaviour in a state of nature and sought its bases in laboratory studies. Anyone who knows something about peck orders in fowls and dominance hierarchies in other vertebrates or has read Lorenz's *On Aggression* will certainly begin to look at human problems from a rather different angle. This is no place to discuss the findings of comparative psychology, but perhaps one observation from the fowl-yard should be quoted. It is simply that altruistic behaviour is never observed except where domination is complete. That is a principle which, appropriately refined to human circumstance, is applicable to most of the situations which keep men awake in the small hours of the morning.

There is only one justification for the intrusion of such ideas on human conflict into a discussion on the natural history of infectious disease. Not so long ago it would have been inconceivable that anyone should attempt to describe disease in terms based on scientific observation and experiment. Even as late as 1880 the most learned historian of disease in England, Creighton, flatly refused to believe in the micro-organismal theory of infectious disease. Today every individual who can read a newspaper has a crude but practically useful background knowledge of microbiology. The housewife preserves fruit by heat sterilization, insists on pasteurized milk and accepts a ritual of boiling the baby's bottle. The soldier accepts his innumerable immunizing injections as a matter of course and is liable to be over-confident that a timely 'shot' of penicillin will eliminate any risk of venereal disease. The essentials have filtered down from the research laboratories through medical practitioners and the press until they have been incorporated reasonably effectively into the social pattern of our times. Similarly, a working knowledge of the bases of nutrition is available to everyone, and as a result our children are the best nourished and probably the healthiest since civilization began.

270

Fever, famine and war are the three traditional scourges of mankind, and the method which has banished the first two is the only one which conceivably could banish the third. An objective scientific sociology drawing its fundamental ideas from the *experimental* study of animal behaviour and analysing social phenomena in the light of these ideas is in the making. Computer techniques have now been perfected which in principle should allow us to extract from observations on human phenomena conclusions as valid as could be looked for if controlled experiment were possible. It is probable indeed that there is already available sufficient factual material to allow a clear picture of the problems of human conflict to be given at the biological level. There must be some way in which an account of that material can be given which is acceptable at the intellectual level to all intelligent people. Some day someone will write a natural history of human conflict built up from a solid experimental basis of work on comparative behaviour in birds and animals and adequate objective non-political surveys and analyses of actual human behaviour. Books of that type could be written with the same biological approach and the same objective honesty as can be found now in a dozen different popular or semi-popular accounts of other aspects of the human situation. Perhaps the diffusion of such ideas would do more than anything else to develop an atmosphere within which a peaceful development of civilization might be possible.

Index

Adenoviruses 54
Aedes aegypti 243
African horse sickness 149
Agammaglobulinaemia 76, 79, 90
   in measles 79
Age incidence of disease 82, 90, 100
   in influenza 97
   in poliomyelitis 82
   in tuberculosis 216
Agglutination, bacterial 86
Air sterilization 109
Algae, blue-green 30
Amoebic dysentery 46
Anaerobic bacteria 39–40
Animals, infections from 112ff
Animal reservoirs 112
Anopheles 233, 236, 237
Anthrax 127
   as BW weapon 266
Antibacterial drugs; antibiotics 173ff
   as chemoprophylactics 180ff
Antibodies
   definition of 71
   diversity of pattern 72, 82
   maternal transfer 98
   production of 73
   role in infections 76, 82
   structure 71
Antigen 72
Antigenic determinants 72
Antigenic drift 159, 210
Antitoxins 78
Arbovirus 54, 69, 113, 248
Asian influenza 108, 210
Australia antigen 253ff
   ecological examples from 10, 11
   epidemiological features 118

BCG vaccine 169, 222
B immunocytes 81
Bacteria 29, 32ff
   agglutination 86
   cell wall 32
   classification 39ff
   ecology 35

fossil 24
   nutrition 29, 34
   phagocytosis 75
   as scavengers 38
Bacterial flora, normal 37ff, 190
Bacteriophages 37, 62, 64
   toxin producing 199
   typing by 63
Bats and rabies 114
Biological warfare 264ff
Birth control 232, 241
Black Death 226
Blood transfusion, infection by 188ff, 254
Budgerigars 18
Bundaberg tragedy 101

Cancer 191
Cell culture of viruses 58, 79
   surface functions 28, 80
Cephalosporins 176
Cerebellar degeneration in kuru 256
   in scrapie 260
Chagas' disease 49
Chemoprophylaxis 110, 162ff, 180, 220
Chick embryo and virus research 58, 208
Chickenpox 154
Child health care 163
Chlamydia 42, 110
   of psittacosis 17
Chloramphenicol 181
Clinical trials 182ff
Clonal selection theory 71
Clostridium 126
Coccidiosis of fowls 11, 50
Cockatoos, psittacosis in 20
Common cold 107, 133
   spread of 133, 134
   viruses of 133
Complement fixation 86
Congenital infections 115
Contraceptive pill 111
Corynebacterium 199
Cowpox 70
Coxsackie viruses 69
Croup 69, 193

273

Index

Crowding and spread of infection 11, 13, 20, 126, 251
Cyclostomes 28
Cytomegalovirus 116, 188
Cytopathic effect (of viruses) 60

DDT 161, 239
DNA, bacterial 36
  plasmid 37
Darwin, illness 49
Delayed hypersensitivity 87, 205ff, 221
Dengue, 112, 248
Diagnostic tests 86
Diphtheria 193ff
  bacillus 194, 199
  carriers 130
  epidemics 127
  history 193ff
  immunization against 170, 198ff
  phage β and toxin 199ff
  toxin, toxoid and antitoxin 171, 195, 197, 200
  virulence changes 198, 201
Domagk 173
Double-blind trials 183
Droplet infection 14, 107, 126
Drug resistance, transmissible 178ff
Dubos 174
Dust, infection by 127
Dysentery
  amoebic 46
  bacillary 107, 178
  in monkeys 47

Echo viruses 60
Ecological sciences 4
Ecology and infectious disease 4, 248
Ecosystems, balanced 138, 143
Electron microscopy 60
Encephalitis 112, 248
  Murray Valley 248
  St Louis 99
English Sweats 134ff
Entamoeba histolytica 46
Enterovirus 106
Environmental sanitation 14, 106, 107
Epidemics 118ff, 129
  exhaustion of susceptibles 129
  explosive 125
  history of 2
  on islands 131
  waves 127, 130
Epidemiology 105ff
Episomes 37, 178

Eradication of disease 155, 158, 173, 224
Escherichia coli 40
Evolution 22, 137
  of animals 25
  of humans 138
  of infection and defence 22ff, 88, 100
  of microorganisms 24, 29
  of protozoa 44ff
  of viruses 65ff

Faroe Islands 132
Fiji 16, 139
Finlay y Barrés 242
Fleming 155
Florey 174
Food chains 7, 9

Genetic factors in resistance to disease 89ff, 137, 142
Genetic information 28
Genetic recombination of viruses 212
Gerbilles 229
German measles, see Rubella
Gonococcus 110
Gonorrhoea 110
Gramicidin 174
Ground squirrels and plague 230

Haemorrhagic fever 112
Heart valve infections 103
Hepatitis, Australia antigen 253ff
  infectious 107, 250ff
  serum 117, 188, 189, 252ff
Heroin addicts 253
Herpes simplex 116, 152, 153
  simian 153
  zoster 154
Herpesvirus 54, 69, 152
Hospital infections 186ff
  staphylococcus 187
Human conflict 267, 269ff

Immunity 70ff
  as epidemiological factor 82ff
  of infection 51
  to influenza 209
  maternal transmission 98
  to measles 80
  to poliomyelitis 95
Immunization 84ff, 162, 164
  by bacterial vaccines 85
  against cholera 85
    diphtheria 170, 197
    influenza 127, 212

Immunization—*cont.*
  measles 167
  plague 84
  pneumonia 76
  polio 96, 167
  smallpox 164
  tetanus 170
  tuberculosis 222
  typhoid 170
  whooping cough 85
  yellow fever 84, 246
Immunocytes 72
Immuno-deficiency disease 76, 90
Immunogen 72
Immunoglobulin 71
  IgA 74, 95, 134
  IgG 71, 77
  IgM 74
Immunological systems, T and B 72, 83, 90
Immunosuppessive drugs 109, 195
Infantile paralysis, *see* Polio
Infectious mononucleosis 69, 111, 188
Influenza 97, 125, 202ff
  in animals 209
  antigenic drift 210
  Asian pandemic 1957 108, 129
  history of 202ff
  hybridization 211
  1918 pandemic 97, 122, 204, 206
  types A and B 206, 208
  vaccines 212
  viruses (*Myxovirus*) 69, 207, 209
Insects, phagocytosis in 27
Interferon 133, 184, 264
Intestinal immunity 167
  infections 158
Intestine, bacteria in 105
  viruses in 106, 167
Isoniazid 220

Jenner 70, 166
Jungle yellow fever 245

Kala azar 49, 112
Koch's postulates 57
Krakatoa 9
Kuru 256ff
  and cannibalism 256
  in chimpanzees 258
  genetic hypothesis 257

Lady-bird beetles 8
Lassa virus 68, 123
Leishmania 49

Leprosy 41ff
Leptospirosis 111
Leukaemia, associated infections 191
Lorenz 270
Louse as vector of typhus 113, 146, 147
Lymphocytes 72, 80
Lymphocytic choriomeningitis 151
Lysogeny 199

Mad itch 149ff
Malaria 50, 112, 232ff
  chemoprophylaxis 162, 240
  development of parasite 50, 233ff
  elimination 158, 237, 241
  history 232
  immunity 136
  in Ceylon 235
  in Mauritius 241
  in monkeys 238
  mosquito transmission 233
  resistant strains 240
Malnutrition and infection 163
Mantoux test 87, 221ff
Marmots and plague 225
Mary, Queen of Scots 203
Mauritius 241
Measles 15, 78, 124
  in adults 121
  encephalitis 168
  epidemics 15, 127
  eradication 158, 169
  in Faroe Islands 132
  in Fiji 16, 139
  giant cell pneumonia 79
  immunity 16
  rash 79, 80
  SSPE 81
  vaccines 167ff
  virus 79
Meningitis 103, 126
  chemoprophylaxis 180
Meningococcus 39, 126, 180
Metchnikoff 26
Mexico, smallpox in 122
Mites 114, 145
Molecular biology 4, 24
Mortality from infectious disease 90
  age incidence 91
  fall in 20th century 1
Mosquitoes 112
  Ades and yellow fever 243ff
  Anopheles and malaria 233ff
  control 161, 239
  myxomatosis, spread by 140

Index

Mouse, multimammate 229
Mouse plagues 11, 146
Murray Valley encephalitis 248
Mycobacteria, atypical 224
Mycoplasma 42
Myxomatosis of rabbits 112, 138ff
  development of genetic resistance 141
  liberation and spread 140
  changes of virulence 140
Myxovirus 54, 69

Nagana 148
Newborn, infections of 187
Newcastle disease of fowls 147
Nucleic acids of viruses 53
Nutrition, animal 6
  of microorganisms 29

Origin of life 22
Orphanage infections 119, 251
Oryctolagus 139, 142
Overpopulation 12, 269
Oysters and hepatitis 252

Pacific islands, infectious disease in 13, 16
Papovavirus 54
Paramyxovirus 54
Parasites and parasitism 7, 11, 54
Pasteur 2, 52, 70, 194
Penicillin 173ff
  chemoprophylactic use 163
  discovery 173
  mode of action 175
  new types 177
  sensitization 181ff
Penicillinase 177
Persisting tolerated infection 252, 259
Phagocytosis 27
Phlebotomus 49
Picornavirus 54, 69
Placebos 183
Plague 14, 112, 225ff
  antibiotic treatment 230
  bacillus 228
  flea transmission 124, 225
  history of 160, 226ff
  immunization against 230
  pneumonic 226
  rodent reservoirs 124, 225
  in South Africa 229
Plagues, viral 60
Plasma cells 73
Pneumococcus 39
  immunization against 76
  specific soluble substance 75

types 74, 76
Pneumonia 74
  in agammaglobulinaemia 76
  treatment by antibiotics 76
Polio (poliomyelitis, infantile paralysis) 91ff
  age incidence 91ff
  'dirty fingers' 94, 107
  epidemics of 92
  history 91
  immunization, natural 93
  Sabin and Salk vaccines 167
  virgin soil epidemics 92, 91
  virus 94, 120
Pomona leptospirosis 111
Population, influence of malaria 232
  pressure 6, 232, 268
  world stabilization of 269
Post-mortem room sepsis 121
Poxvirus 54
Progress in 1970s 263
Protozoa 212ff
Psittacosis 17, 99, 109
  in Australian parrots 18, 20
  control of 161
  in Louisiana 123
Puerperal fever 101, 122, 186

Q fever rickettsia 120
Quarantine 2, 109, 159ff
  of animal disease 160
Quinine 240

RNA, double stranded 78
RNA, infectious 62
Rabbits, myxomatosis 138
Rabies 69, 114
  bat-borne 114, 124
Rafflesia 66
Rat control 228
Rats and plagues 228
Receptors, immune 72
Relapsing fever 112
Reservoirs, animal 112, 124
Respiratory infections 96
  age incidence of death 97
  in isolated communities 133
Respiratory syncytial virus 98, 125
Rhabdovirus 54, 69
Rheumatic fever, chemoprophylaxis 163
Rhinoviruses 133, 134
Rickettsiae 42, 65, 143ff
  evolution of 65, 147
Rinderpest 151
Rockefeller Foundation 244

Rocky Mountain spotted fever 112, 143
    mode of spread 144
Rubella 115
    foetal infection 115

Sabin vaccine 85, 167
Sacculina 66
Salk vaccine 85
Salmonella 40
'Salvarsan jaundice' 252
Scale insects 8
Scarlet fever 108, 200
Schick test 86, 195ff
Scrapie 260ff
Scrub typhus 112, 145
Self and not-self 26, 72
Semmelweiss 122, 186
Shakespeare 152
'Shedding' of microorganisms 122, 125
Sleeping sickness 48, 112, 148
Slow viruses 256, 261
Smallpox 70
    epidemic spread 122, 123
    eradication 158
    vaccination 70, 166
Spirochaetes 41
Spitzbergen 132
Spores, bacterial 126
Standard of living 95
Staphylococcus 38, 101, 187
    drug resistance 119, 176ff, 187
    pneumonia 121
'Stranger's cold' 133
Streptobacillus moniliförmis 10
Streptococcus 39, 101, 103, 108, 121
    chemoprophylaxis 181
    and scarlet fever 200
Streptomycin 174, 220
Subacute bacterial endocarditis 103, 181
Subclinical infection 82, 84, 118, 163, 195, 251
Sulphonamide drugs 76, 173
Surgical infections 187
Susceptibility to disease 88ff
Swine influenza 209
    fever 149
Sylvilagus 139
Syphilis 41, 110
    congenital 115

TAB vaccine 85, 107
Tetanus 77
    toxoid immunization 84, 171
Tetracyclines 176
Threadworms 107

Thymus dependent (T) cells 77, 80
T immune system 77, 79, 99
Ticks 144
Tobacco mosaic virus 62
Tolerance, immunological 72
Toxoids 85, 171
Toxoplasma 116
Trachoma 111
Transovarial transmission 148
Transplantation of tissues 114
Triple antigen 171
Trypanosomes (*see also* Sleeping sickness) 44,
    148ff
    evolution 148
    transmission 48, 148
    in wild animals 50, 148
Tsetse fly 48, 148
Tubercle bacillus 213
Tuberculin 213
Tuberculosis 213ff
    bovine 214
    control by BCG 222ff
    drug treatment 180, 222
    genetic influence 216
    Mantoux test 77, 221
    mass radiography 223
    primary infection 214
    reduction since 1950 218ff
    sensitization 77, 215
    spread 128
    in susceptible races 218, 219
Twins, concordance in 89
    fraternal and identical 89, 216
    incidence of TB in 216
Typhoid bacillus 32
    carriers 126
    phage typing 63
Typhoid fever 106
    control 160
Typhus fever 145ff
    louse borne 112
    murine 12, 112, 146
    in Naples 1942 161

Urinary infections 187

Vaccination, Jennerian 70, 85, 166ff
Vaccines, bacterial 85, 170
    killed 170
    live virus 166ff, 246
Vampire bat 114
Varicella 154
Variolation 164
Vectors, arthropod 112
Vegetation, climax state 9

Index

Venereal disease 110
Vertical transmission 115
Virgin soil epidemics 138, 142
Virions 53
Virus disease of man 54
Viruses 52ff
    bacterial (*see* Bacteriophages)
    classification 69
    evolution 65
    history of research 52
    neutralization by antibody 86
    variation and genetic manipulation 267
Viscerotome 246
Vital statistics 127, 241

War, influence on infectious disease 147, 164
    nuclear 264

Warfare, biological 265
Web of life 138
Weil's disease 111
Whooping cough vaccine 170
Wound infections 120
    by *Pseudomonas* 187

Yellow fever 112, 244ff
    antibody surveys 244
    control 246
    epidemiology 136, 243, 245
    history and spread 142, 242
    immunization 85, 244, 246, 252
    infection and immunity 244
    jungle 245
    virus 243